W9-ACQ-912

Empowering Technology

Empowering Technology

Implementing a U.S. Strategy

edited by Lewis M. Branscomb

with contributions by
Lewis M. Branscomb
Harvey Brooks
Brian Kahin
George Parker
Gene R. Simons
Dorothy S. Zinberg

The MIT Press, Cambridge, Massachusetts, and London, England

©1993 Massachusetts Institute of Technology and the President and Fellows of
Harvard College

All rights reserved. No part of this book may be reproduced in any form or by any
electronic or mechanical means (including information storage and retrieval) without
permission in writing from the publisher.

This book was set in New Baskerville at The MIT Press and was printed and bound in
the United States of America.

Library of Congress Cataloging-in-Publication Data

Empowering technology : implementing a U.S. strategy / edited by Lewis
    M. Branscomb ; with contributions by Lewis M. Branscomb . . . [et
    al.].
        p.   cm.
    Includes bibliographical references and index.
    ISBN 0-262-02366-0 (hc). — ISBN 0-262-52185-7 (pb)
    1. Technology and state—United States.   2. Industry and state—
United States.   I. Branscomb, Lewis M., 1926–
T21.E53   1993
338.97307—dc20   93-23975
                                                          CIP

Contents

Preface

The research program on dual-use technology carried out in the Science, Technology, and Public Policy Program (STPP) of Harvard's John F. Kennedy School of Government between 1987 and 1991 led to the publication in May 1992 of *Beyond Spinoff: Military and Commercial Technologies in a Changing World.*[1] That book explored the complex relationships between government-funded and commercially-funded technology and concluded that the *de facto* technology policies that served U.S. security needs so well in the four decades after World War II would have to be supplemented by new public-private relationships to serve both defense needs and commercial competitiveness. The final chapter of *Beyond Spinoff* sketched a framework for a new relationship between government agencies and commercial industry.

Empowering Technology takes *Beyond Spinoff* as a point of departure and examines seven of the major technology policy issues the United States is struggling with in 1993 with economic revitalization as the nation's priority goal. Defining the right objectives for the government's role in the nation's science and technology is relatively straightforward. This book focuses, therefore, on the conditions for successful implementation of policy. Two ideas emerge as central: first, the United States must shift its strategies from those centered on government to strategies centered on industry. Second, we must balance government investments in technology creation with a new focus on promoting the utilization of technology.

viii
Preface

We conclude that only if the business community shares with government a new relationship of mutual trust will the vision be realized. Thus we commend the ideas in this book to business leaders and government officials in Congress and the executive branch, as well as scientists and technologists.

The authors express their gratitude to the Alfred P. Sloan Foundation for its support of the project entitled "The United States Science and Technology System in its Global Context" (Grant 90-10-1) on which this book is based. Additional assistance from Mitre Corporation and from the group of companies (IBM, DEC, Apple, AT&T, Bellcore, and McGraw-Hill) supporting the Information Infrastructure Project is gratefully acknowledged.

The ideas in this book owe much to participants in the STPP Information and Innovation Working Group in the Kennedy School Center for Science and International Affairs. Participants have included, in addition to the authors of *Beyond Spinoff* and the authors of this book: David Allen, David Guston, David Hart, Donald Hornig, Megan Jones, Donald Kerr, and Charles Zraket. David Guston made particularly important contributions to Chapters 4 and 7, as well as other sections of the book. Christofer Tucci, doctoral student at MIT, was also helpful in the bibliographic research. Brigette Pak provided much-appreciated research support.

A very special thanks is owed to Teresa Johnson, CSIA's Senior Editor, to Stephen Stillwell, CSIA librarian, and to Nora Hickey, Miriam Avins, Helen Snively, and Karen Shepard, whose dedication to completing this project on time went far beyond normal expectations. Teresa Johnson not only greatly improved the book's clarity and readability, but guided its development under a killer schedule. We also thank Larry Cohen of MIT Press for his part in bringing the book out so quickly.

Lewis M. Branscomb
Cambridge, Massachusetts
May 8, 1993

Note

1. John A. Alic, Lewis M. Branscomb, Harvey Brooks, Ashton B. Carter, and Gerald L. Epstein, *Beyond Spinoff: Military and Commercial Technologies in a Changing World* (Boston: Harvard Business School Press, 1992).

About the Authors

Lewis M. Branscomb is Albert Pratt Public Service Professor at the John F. Kennedy School of Government of Harvard University, where he directs the Science, Technology, and Public Policy Program in the Center for Science and International Affairs (CSIA). A physicist educated at Duke and Harvard Universities, Branscomb has served as director of the National Bureau of Standards (now the National Institute for Standards and Technology [NIST]), and as vice president and chief scientist of IBM Corporation. In 1979 Dr. Branscomb was appointed to the National Science Board, and served as Chairman for four years. A member of the Carnegie Commission on Science, Technology and Government, and an author of several of its reports, he was also principal investigator of the dual-use technology research described in *Beyond Spinoff: Military and Commercial Technologies in a Changing World*, and was one of its authors. He is principal investigator of the Sloan Foundation supported research program on which this book is based.

Harvey Brooks is Professor of Technology and Public Policy (Emeritus) in the Center for Science and International Affairs of the Kennedy School of Government, Harvard University. He received his PhD in physics from Harvard in 1940, served as Dean of the Division of Engineering and Applied Physics at Harvard from 1957 to 1975, and was Gordon McKay Professor of Applied Physics from 1950 to 1982. He served on the President's Science Advisory Committee from 1957 to 1964, and as a member of the National Science Board from 1962 to 1974. He was chairman of the Committee on

Science and Public Policy (COSPUP) of the National Academy of Sciences from 1966 to 1971. He directed the Science, Technology, and Public Policy Program of the Kennedy School from 1975 to 1986. Most recently he co-chaired with Dr. John Foster the Committee on Technology Policy Options in a Global Economy of the National Academy of Engineering, whose report, "Mastering a New Role: Shaping Technology Policy for National Economic Performance," was released in March 1993.

Brian Kahin directs the Information Infrastructure Project in the Science, Technology and Public Policy Program at Harvard University's John F. Kennedy School of Government. He recently edited *Building Information Infrastructure* (McGraw-Hill/Primis, 1992), a book on issues in the development of the National Research and Education Network. A graduate of Harvard College and Harvard Law School, Mr. Kahin is General Counsel for the Interactive Multimedia Association, based in Annapolis, Maryland; he also directs the association's Intellectual Property Project.

George Parker is a PhD candidate in political science at the Massachusetts Institute of Technology. His major research interest is U.S. science and technology policy, with a special emphasis on the problems of technology transfer.

Gene R. Simons is an Industrial Engineering Professor in Decision Sciences and Engineering Systems at Rensselaer Polytechnic Institute, where he founded and from 1989 to 1992 directed the Northeast Manufacturing Technology Center. He chaired the Industrial and Management Engineering Department at Rensselaer from 1971 to 1986. In 1992 he was a Senior Research Fellow in the Science, Technology, and Public Policy program at CSIA in Harvard's Kennedy School of Government. He is a fellow of the Institute of Industrial Engineers and has received the Darrin Counseling Award and the LEAD Award from the Society of Manufacturing Engineers. Dr. Simons received his bachelor's degree in Management Engineering from Rensselaer in 1957 and, after nine years in industry, returned to Rensselaer where he received his PhD in Management Science in 1969. He is known for his work in the management of technical projects and its relationship to manufacturing policy, technology transfer, and new product development. Dr. Simons is currently working with NIST to redesign the federal system of industrial extension.

Dorothy S. Zinberg is Lecturer in Public Policy and a Senior Research Associate in the Center for Science and International Affairs at the Kennedy School of Government, Harvard University. For ten years a biochemist at Harvard Medical School, she later received a PhD in sociology from Harvard. Since then she has carried out research in the social science of science, focusing on human resources in science, engineering, and technology. She has been a member of the Board of International Scientific Exchange at the National Academy of Sciences (NAS), the NAS Advisory Board of the Council for the International Exchange of Scholars, and its Commission for International Relations. She chaired the advisory committee to the International Division of the National Science Foundation, and served on the Congressional Office of Technology Assessment Advisory Panel, "Sustaining the National Technological Base: Education and Employment of Scientists and Engineers." At the present time she is a member of the National Research Council's Board on the Future of Engineering Education. She has written extensively on the international mobility and education of scientists and engineers; most recently she co-authored and edited *The Changing University: How the Need for Scientists and Technology is Transforming Academic Institutions Internationally* (Kluwer Academic Publishers, 1990).

1

The National Technology Policy Debate

Lewis M. Branscomb

In a world in which comparative economic advantage derives principally from development and use of intellectual resources manifested in new technology-based products and services, we are perhaps better positioned than any nation in the world to maintain and increase our comparative economic advantage. The policies and programs now being considered are a major step in the direction of harnessing the incomparable intellectual resources that reside in our universities, industry, and government to reverse the trend toward diminishing world market share of United States–based production.

Robert M. White, President, National Academy of Engineering, Testimony to the Senate Committee on Commerce, Science and Transportation, March 25, 1993.

The 1992 United States elections brought to a head a national debate about United States technology policy that has been building since the early 1970s.[1] All of the candidates for the presidency called attention to profound changes in American economic, security, and environmental circumstances. Along with growing global interdependence, these changes have evoked a call for changes in government policy.[2] In February, 1993, the Clinton-Gore administration issued its ideas for a comprehensive technology policy to enhance United States industrial performance in world markets.[3] These policies are part of a sharp acceleration of federal activity in support of industrial competitiveness and other domestic goals. Their pace and scale, and the level of confidence in their effectiveness, differ dramatically from initiatives that had already been put in place by the Bush administration and the Congress.[4]

The Clinton program presents Congress with a daunting agenda of proposals, most of which require new legislation for their implementation. Candidate Clinton promised the American voters he would change the government and its policies. There is abundant evidence that as president he will try to make good on this promise, but implementing it will not be easy. The federal agencies will find the new tasks a poor fit to their existing structure and their experience developed over forty years of Cold War. Moreover, the fifty states will find that they are expected to play a major role not only in executing many of the new initiatives, but in contributing to their financial support as well. They will not find it easy to assure the stability of funding for the new activities, and they may need new institutional mechanisms to harmonize their activities and to negotiate with the federal government with one voice rather than fifty.

The election of President Clinton and Vice President Gore brought a new generation to power in the United States federal government, the first elected to these offices since World War II who did not participate in that war. No longer are United States views on international security and foreign affairs dominated by super-power competition with a nuclear-equipped and sometimes expansionist Soviet Union. The end of the Cold War, the rise of economic competition in an increasingly global economy, and rapid change in the nature and power of science and engineering give the United States a new political, economic, and security agenda, but leave it with most of the same governmental structures that were built during the Cold War. Many Cold War–era concepts about the nature of technological innovation, and the part science and engineering play in it, still dominate political thinking.

Thus many of the concepts and new policy initiatives that have appeared in the late 1980s and early 1990s are transitional. This is also true of the missions of federal agencies (especially Defense and Commerce), the roles of national institutions such as government laboratories, universities, and consortia of firms, and especially of the role of state governments in the promotion of economic activity through investments in technology, in industrial extension services, and in human resources. This chapter identifies the postwar origins of United States science and technology (S&T) policy and the institutions that drive it, analyzes the elements of policy that are in flux, and outlines the issues that are most likely to be hotly debated in the Clinton administration's first years. These issues are explored in the chapters that follow.

Thinking about Technology Policy

A technology is the aggregation of capabilities, facilities, skills, knowledge, and organization required to successfully create a useful service or product. Technology policy concerns the public means for nurturing those capabilities and optimizing their applications in the service of national goals and the public interest. Technologies are created for economic reasons and the investments they call for must be economically justified. Technology is an important element of industrial success, but it is only one element, along with labor productivity, capital cost, and managerial skill. Technologies are almost never an end in themselves.

The boundaries that distinguish technology policy from economic and industrial policy are fuzzy at best. A sound macroeconomic environment is essential for innovation and technical progress, but it is not sufficient. Market forces and the competition they engender are the primary incentives for the creation and use of technology. Market failures abound in technology generation and use, however, and constitute the primary justification for government efforts to empower technology to contribute to economic health. The traditional gulf in perceptions between macro-economists and technologists has been an obstacle to public policy consensus in the past.

Technology policy must include not only science policy—concern for the health and effectiveness of the research enterprise—but also all other elements of the innovation process, including design, development, and manufacturing, and the infrastructure, organization, and human resources on which they depend. There is widespread agreement that the government's role is to enhance the competitive advantage of United States firms in international commerce and to increase innovation rates and productivity here at home, without disrupting markets or spending public funds inappropriately.

The administration's technology policy comprises three primary elements:

•Supply-side activities: federal and state research and development (R&D) activities to create new technology which can contribute both to government missions and to private sector innovation and productivity;

•Demand-side activities: federal-state cooperation to improve availability, adaptation, and exploitation of technology by small to medium-sized firms, and activities to encourage investments in technology, including investments in education and training; and

•Information and facilities infrastructure: networks of people and digital communications to create markets for information services, enhance linkages between firms and promote cooperation between all economic sectors, public and private, and create specialized, shared technical facilities, such as those for materials testing, design simulation, and microelectronics circuit foundries.

The Bush and Reagan administrations largely confined themselves to the first of these activities: creating technology in pursuit of federal missions. The Clinton administration's strategy embraces all three with apparently equal enthusiasm.

The design of a national technology policy must be predicated on the reality that it is the private sector, not the government, that creates wealth using science and technology. The technology the private sector uses is largely created by private firms, only supplemented by "spin-off" from federal agency mission-driven technology and "trickle down" from high science. In a capitalist society wealth is created by enterprises, and is then shared among employees, owners, customers and business partners. Thus public investment in technology for commercial benefit is a "trickle down" policy by its nature. It unavoidably involves the conversion of public assets to private benefit. The legitimacy and accountability of government in the administration of a technology policy is, therefore, a serious concern. This reality places a major political constraint on both policy and institutionalization of policy innovations in this area. These constraints are discussed in Chapter 9.

To be politically stable, the benefits of government activities should be seen to accrue substantially and preferentially to United

States firms and workers; otherwise political support will erode. Activities in pursuit of technology policy must also be capable of competent administration without capture and distortion by benefitted constituencies. They must also be informed by competent advice from the private sector.

United States S&T policy is, with the exceptions noted below, largely uncodified; it must be deduced from the laws and organization of government and by observation of the actions of government managers and agencies. A wise veteran of Washington politics once said that, "in Washington the budget *is* the policy." That policy is continuously in flux, and it is unclear what direction *de facto* policy will take in the next decade. But certain generalizations about the directions of policy change can be made with some confidence:

•The federal government is shifting its priorities away from R&D in pursuit of government missions in defense, space, and atomic energy toward technology development in collaboration with private industry to support enhanced industrial competitiveness.

•Government technology policy is shifting from almost exclusive emphasis on generation of new technology, to increased support for helping both individuals and firms obtain, adapt, and use technology. This shift to a "demand-side" technology policy involves major investments in information infrastructure, in training and education of the workforce, and in industrial extension services.

•Private firms, in a more relaxed anti-trust enforcement environment and responding to international competitive pressures and to changes in innovation processes, are moving away from vertical integration toward alliances of many kinds with suppliers and distributors, other firms, and members of industrial consortia for pre-competitive and infrastructural R&D.[5]

•Government decisions on civil technology activities will be increasingly made in collaboration with private-sector bodies, and programs will be undertaken in evolving forms of public/private partnerships. This will require some loosening of the barriers to private-sector participation in bodies advising government, and government agencies will be called on by the Congress to "benchmark" the technological standing of the sectors of United States manufacturing and service industries, relative to that of their foreign competitors.

•All of these activities place heavy burdens on state governments, which must become increasingly sophisticated in the services they provide to small and medium-sized businesses to improve productivity and attract capital investment. The federal government will have to develop more effective relationships with the states, as agencies expand their support of state activities.

•After a difficult and protracted period of downsizing, the defense effort (both procurement and R&D investment), while still an important driving force in United States technology, will no longer be the dominant force it has been in the last forty-five years. Defense will begin to acquire increased fractions of its technology from commercial sources (foreign as well as domestic), so that per dollar spent, defense's contribution to competitiveness may actually grow.

•The national laboratories, which in 1992 performed twice as much federally-funded R&D as did universities, are struggling to find new missions to sustain their current levels of effort; failing that, they will face severe downsizing, despite strong political support from the affected communities. This downsizing may be delayed by the political power of constituencies supporting them, and by national respect for the laboratories' skills, but it will come.

•The research universities, a major source of new scientific and engineering ideas, will be increasingly hard-pressed, financially and politically. They are losing some of the autonomy they have enjoyed in choosing research objectives and determining which investigators will be funded to pursue them. They are becoming more deeply engaged with creating "useful" knowledge and accelerating its diffusion to the private sector. Nevertheless, they will continue to focus primarily on basic research that is open to all.

•International cooperation will increasingly influence American science and engineering. Many scientific facilities, such as the Supercollider accelerator, and engineering projects such as the Space Station, are becoming too expensive for any one nation to pursue unaided. In the future such projects are more likely to be multinationally designed and sponsored. Governments feeling pressure to reduce trade frictions are beginning to internationalize some of their civilian technology development activities, such as the Japanese Intelligent Manufacturing Systems project. The shift of publicly sponsored R&D from military to industrially useful technology will collide, at times, with responses to continued trade frictions, and thus intellectual property policy, export controls, and policies toward international projects will remain in the political limelight.

Each of the above trends will present new political difficulties in defining the line between public and private responsibilities, and in measuring the benefits from federal technology programs. The new programs will have to be very carefully thought out, with the government role precisely defined and delimited, in order to achieve a politically acceptable level of confidence in the fairness and effectiveness of government activities.

The United States is moving toward a more integrated industrial technology base. That industrial base is increasingly integrated into the world-wide economy. That integration suggests that there will be a convergence of the policy tools for guiding technology in all three areas: military, commercial, and environmental. No longer will defense technology be generated solely in a command economy; instead defense agencies will invest in dual-use technology and buy more commercially-available products. Economic competitiveness will no longer be left to a *laissez-faire* economic policy; government will share costs of base technology development with commercial firms. Investments in environmentally superior technology, today forced by regulatory pressures, will receive increasing direct public support.

The basic law that defines United States technology policy is the National Science and Technology Policy, Organization, and Priorities Act of 1976.[6] The occasion for this statute was the legislative establishment of the Office of Science and Technology Policy (OSTP) in the Executive Office of the President during the administration of President Ford. In January, 1973, President Nixon had reorganized and down-graded the predecessor agency, the Office of Science and Technology (OST), by assigning its functions to the National Science Foundation, effectively removing it from the White House.[7] The 1976 Act for the first time articulated the government's role in technology, and subsequent legislation about critical technologies or technological priorities often refers to it. The Act gives policy criteria for federal investments in technology for purposes other than established government missions:

> It is further an appropriate Federal function to support scientific and technological efforts which are expected to provide results beneficial to the public but which the private sector may be unwilling or unable to support[8]....Explicit criteria, including cost-benefit principles where practicable, should be developed to identify the kinds of applied research and technology programs that are appropriate for Federal funding support and to determine the extent of such support.[9]

The *de facto* United States technology policy is, indeed, shifting rapidly from its emphasis on development of technologies to serve missions assigned to the federal government, particularly in the defense, space, and nuclear fields, to programs intended to enhance the economic performance of private industry. The general directions for this policy evolution were laid out by Presidential Science Adviser D. Allan Bromley in 1990 in the first formal statement of technology policy by the Bush administration.[10] The year before, President Bush had publicly declared his intention to provide cost-shared support for "pre-competitive, generic" technologies of value to commercial industry, as Congress had authorized in the 1988 Omnibus Trade and Competitiveness Act.[11]

During 1992, perhaps prompted by the presidential election campaign, the Bush administration substantially accelerated its initiatives in technology support to the economy, even as its political rhetoric continued to warn against industrial policy. It requested growth in the Advanced Technology Program in the Department of Commerce to $69 million for Fiscal Year 1993. It announced new R&D initiatives in High Performance Computing and Communications, including additional investments in the National Research and Education Network (NREN).[12] Interagency initiatives in Advanced Materials and Processing and in the non-defense component of manufacturing research and development were planned. Nevertheless the Republican administration remained staunchly ideologically committed to *laissez-faire*. Any hint of "industrial policy," "picking winners and losers," or strategic planning for the economy were castigated as harmful to private industry and to the economy. More often than not, in Bush administration rhetoric, technology policy was seen as indistinguishable from the despised "industrial policy," notwithstanding publication by the Bush White House of its own *United States Technology Policy* in September 1990.

President Clinton has reversed the Reagan-Bush ideological bias against technology policy. His administration has adopted a centrist policy that emphasizes services to small business, national information infrastructure based on private commercial services,[13] and federal investments in technologies in which private firms are likely to underinvest. The administration's technology strategy was anticipated in the September 21, 1992, statement of the Clinton-Gore presidential campaign organization,[14] and is embodied in their policy declaration of February 22, 1993.[15] These policies call for a sharp increase in federal support for industrial competitiveness. Nevertheless, there is still uncertainty in the nation as a whole about the appropriate role for government funding of science and technology in support of economic performance.

After the new administration's first 100 days, it is still unclear how enthusiastic the Clinton administration will be for technology investments as a tool for economic revitalization, or how strong the voice of advocates of technology policy will be on the National Economic Council. Great prominence has been given to the "economic team" of senior officials in the new administration, but macroeconomic issues, such as reducing the deficit, increasing investment, and reducing costs of health care, overshadow the longer-range investments in technology and its diffusion.

Science and technology policy will receive high-level attention, however, from the vice president, who has been given oversight of science and technology issues and of "reinventing government" to gain efficiency and effectiveness in agency operations. The president's advisor on science and technology, Dr. John Gibbons (former director of the Congressional Office of Technology Assessment) works closely with the vice president.[16] He also directs the Office of Science and Technology Policy (OSTP), a statutory agency in the Executive Office of the President. His office is the focal point for technology policy, which will gain strength from the vice president's political influence.

The biggest challenge facing the Clinton administration's "vision for a changed America" may not be finding public and political support for change, but rather the inertia of institutional structures and policy paradigms developed during the Cold War. To understand how very different the new approach must be from the old, we should consider where the current policies and structures came from.

Postwar Science and Technology Policies[17]

During the four decades after the Second World War, the United States attained the highest level of scientific and technological achievement in history. New industries with revenues of hundreds of billions of dollars were created from scratch after the war. Stimulated by investments for defense, space and atomic energy and the creative powers of American science and engineering, the aviation, computer, and microelectronics industries became commercial leaders throughout the world. American universities attracted the interest and admiration of all countries, and they still attract one-third of all the students in the world who study abroad. In only five years, Americans created the organization, the facilities, and the technology for manned exploration of the surface of the moon.

The policies under which these achievements occurred were articulated in Vannevar Bush's report to President Truman, entitled *Science—the Endless Frontier*.[18] The period from the early 1950s to about 1968 (when the growth of American science came to a halt for ten years) is often called the "golden age of American science."

Postwar federal policy for science and technology had two parts: government support for research in basic science, and active development of advanced technology by federal agencies in pursuit of their statutory missions.[19] Dr. Henry Ergas characterizes this policy, like that of Britain and France, as "mission-oriented" technology policy.[20] He contrasts this approach with the "diffusion-oriented" policies of Germany, Switzerland, and Sweden.

United States S&T policy since World War II has been based on five assumptions about the process of innovation and the role of government in attempting to foster it. These assumptions served

well enough during much of that period, but have become outdated, and in some instances have proved to be unsupported. Their
shortcomings will be explored in this chapter. The five assumptions
are:

•Basic scientific research is a public good. Investment in basic
science, especially in combination with higher education, leads
through a sequential process of innovation to the creation of new
technologies which in turn may spawn new industries. A "social
contract" between government and science accords to the scientific community a high degree of autonomy in selecting targets for
basic research and construction of competitive, merit-based processes for selecting projects for public support, in return for a
political consensus that science, pursued this way, will provide the
nation benefits far in excess of their cost.

•In fulfillment of the government's responsibilities for defense,
space exploration, and other statutory responsibilities, federal
agencies should aggressively pursue the development of new technology for use in these missions. The technological fruits of such a
mission-driven high-technology strategy will automatically, and
without cost to the government, "spin off" to commercial uses, thus
further stimulating industrial innovation.

•Reliance on market forces to stimulate industrial competitiveness
is not compromised so long as the government refrains from direct
investments in research to create technology specifically for commercial exploitation, and leaves to private industry the responsibility for tapping into government sources of science and of spinoff
technology.

•With the generally accepted idea that environmental costs have
only negative impacts on the economy (which is misleading in that
it fails to reflect a future world market in environmentally useful
technologies), the government has relied on regulations to force
private investment in environmentally useful technology.

•The United States government has viewed science and technology
as assets to be deployed internationally in support of the political
goal of strengthening alliances to contain the Soviet Union. The
"peaceful uses of atomic energy" program and the Landsat program of earth observations from satellites were designed to make

these military-driven technologies more acceptable to publics at home and abroad.

Each of these elements of postwar American policy entailed a tacit assumption about the mechanism through which government investments in research and development would contribute to industrial innovation, and hence to the competitiveness of American products in the world market. These assumptions were derived from a "supply-side" picture of how the process of innovation works in an industry based on high technology. The United States became the world technology leader by virtue of its enormous government-funded defense, space, and energy activities and associated basic science investments. It followed that preserving a perceived comparative security advantage would require that technology exports should be controlled, despite the fact that technology diffusion abroad might serve U.S. commercial and political interests.

Today, the national security argument for strict export controls on technology is greatly weakened, and Americans appreciate that there are other nations at essentially the same level of technological capability. That equality, however, is not driving Americans away from governmental activism in commercial technology; rather the opposite. The United States government is rapidly expanding its publicly funded investments in civilian technologies and in technology diffusion. The danger is that as government investments in civilian technology grow, Congressional calls for constraints on its diffusion abroad through commercial channels may grow as well.

Dominance of Defense R&D

World War II, and the Cold War that followed, dominated government S&T strategy in the postwar period and gave it its "supply-side" or "mission-oriented" character. In 1960, U.S. defense R&D comprised a third of all the R&D, public and private, performed in all the OECD countries.[21] U.S. military R&D was, in effect, the dominant engine in the non-communist world for technological development of the emerging "high-tech" industries. Military procurement and government-funded R&D were big factors in the early postwar development of the U.S. electronics, computer, and aircraft industries.[22] Even if the process of diffusion of military technology to commercial firms was slow, foreign firms, recovering from the war, were unable to challenge the U.S. industrial lead in these markets. American universities were prolific sources of new science from which technologies evolved.

In contrast with today's environment of Congressional distrust and confrontation in defense acquisition, in the 1950s and 1960s defense agencies could take more technical risks and enjoyed a healthier partnership with their contractors. Much of the stimulation given by defense to technology came through adventurous procurement, not through funding of R&D. A massive national science and technology enterprise was built, with many institutional innovations. Those postwar institutions and relationships shaped the policies we inherit today. A system of government-funded national laboratories emerged to support Cold-War technology needs. Today these laboratories enjoy a major share of federal research dollars ($20.8 billion in 1992). The national labs must deal with uncertainties about their missions and weakened relationships with the private commercial sector. Unfortunately they remain heavily constrained in form and structure and are difficult to change.

Industry finds itself divided into two weakly coupled economies—one of defense companies, the other serving civilian markets. Although most of the large prime contractors are subsidiaries of much larger commercial organizations, the barriers between military and commercial units effectively discourage the sharing of technology between them. (See Figure 1-1.)[23]

"Pipeline" and "Spinoff": Images of the Contribution of Science to Technology

Bipartisan Congressional support for science has rested heavily on acceptance of the "pipeline" model of the process by which social returns arise in the form of industrial innovations.[24] This conventional, somewhat discredited, model assumes that innovations arise in the research laboratory or the inventor's shop, and after a sequence of discrete steps through applied research, development, and design, are produced and marketed. It is further assumed that this process is more or less automatic and inevitable. A science policy oriented to basic research is, in this model, the starting point for innovation; a government commitment to funding basic research would, under this assumption, constitute a successful technology policy.

The "pipeline" model is not a bad description of how new industries arise from new science—a process that takes a decade or more. The model served the United States well during the postwar period of reconstruction when the United States had no serious overseas competition in the field of advanced technology. The emerging bio-technology industry, for example, owes its origins primarily to research in molecular biology, genetics, and biomedical science funded by the National Institutes of Health. (In the last decade, NIH has invested over $60 billion in research conducted in universities and the NIH institutes.) The pipeline model is, however, inapplicable to the way established industries compete through rapid incremental progress, and it is an even less appropriate description of how firms in high technology compete in the 1990s.[25] Most commercial innovations are driven by market opportunity, not by scientific discovery.

The second premise of American postwar policy is that the technology created in pursuit of governmental missions will automatically flow to industry and thus contribute to prosperity. The process through which this is presumed to happen is called "spinoff."[26] A key reason for its appeal is that spinoff, like the "pipeline" from basic science to innovation, is assumed to be automatic and cost-free to the government. Both of these assumptions, drawn from "supply-side" economic ideas, have the attractive feature that, if spinoff is automatic and cost-free, the government does not have to select or subsidize the firms or industries that will benefit from commercializing government R&D (referred to by conservatives as "picking winners and losers"). Government can claim that its policies achieve the goals of economic growth without interfering with the autonomy of private firms.

As a further guarantee against central political control over scientific and engineering activities, American policy after the war called for a highly decentralized responsibility for investing in research and development by federal agencies. Under principles advanced by Vannevar Bush and implemented by President Eisenhower following the Steelman Report a few years later,[27] all federal agencies were to develop the technology needed for their assigned tasks, and were also to support a proportionate share of the country's basic research as a kind of "mission overhead" reinvestment in the basic knowledge on which their technology depended. Governmental support of basic research is justified, economists say, because the social returns from basic research exceed its cost. Private returns to a firm investing in basic science are often less than its costs because of low appropriability of the benefits (an instance of "market failure"). Edwin Mansfield's analysis of all the evidence suggests that the social return from basic science in the United States in the late 1980s was about 28 percent.[28]

The Shift from Military to Civil R&D

The Clinton-Gore policy calls for three major shifts from the Cold War period: from defense to commercial industry, from technology creation to information infrastructure and services to disseminate and utilize technology, and from programs executed by federal agencies to those carried out by state governments and commercial firms. The policy is activist, in that governments—both federal and state—are given a bigger role in commercial technology and its diffusion. But it is also a passive policy, in that almost all the proposed program activity depends on initiatives from outside the federal government. In this latter sense, the policy is much less intrusive than the "command economy" of traditional defense, space, and energy R&D.

Some aspects of the strategy are exceptions to this industry-centered approach. The most obvious is the administration's conviction that Cooperative Research and Development Agreements (CRADAs) between the national and federal laboratories and commercial firms will be an effective way to provide support to the industrial technology base without distorting markets or constraining competition. (This issue is addressed in Chapter 4.) The administration also identified three industries for special support: computers and communications, automobiles, and high-speed rail. In each of these cases there is a public-good dimension to the choice that serves as justification: information infrastructure, cars that pollute less and consume less energy, and advances in public transportation.

The administration proposes a dramatic shift in federal R&D priorities. A shift of $8.7 billion in R&D from defense to civil technology is to be accomplished by 1998. This shift will change the distribution of federal R&D from 41 percent civil and 59 percent military, to roughly equal civil and military contributions.[29] In addition, the Defense Advanced Research Projects Agency (DARPA) has been renamed ARPA, with a strong emphasis on the support of "dual-use" technology.[30]

Both defense and commercial technologies grow from the same seeds of scientific knowledge, benefit from similar tools, techniques, processes and materials, and draw on the same system of education for scientists, engineers and mathematicians. Technologies may be shared by multiple applications, and technological knowledge may be diffused from one domain of use to another. Successful applications of knowledge received through this diffusion process may inspire new uses by the originator of the knowledge. Thus the processes through which technological synergy may be enjoyed are complex and often unseen and undocumented. They are no less important for being obscure.

However, the differences between civil and military innovation cultures are substantial, and call for different relationships between government and industry. Figure 1-1 suggests some of these differences.

Government R&D institutions such as those in the National Aeronautics and Space Administration and the Department of Energy partake more of military than civilian industrial culture. Their R&D derives from government missions in which performance and function are the goal, rather than low cost or rapid incremental improvement. Government missions are fixed by Congress and change only slowly. The resulting activities are usually out of touch with the fleet-footed, low-cost, high-quality manufacturing so vital to manufacturers of commercial products.

Megaprojects

The prototype of mission-centered technology generation is the "megaproject," which can be loosely defined as a unique, usually technically heroic, federally-funded project of high public visibility

Figure 1-1 Two Cultures: Civil and Military Innovation

Attribute	Civil	Military
Design-driven	Driven by markets	Driven by "requirements"
Innovation style	Incremental	"Big leaps"
R&D Intensity	Moderate	High: 4.9 times civil firms
Product cycle	Measured in years	Measured in decades
Technology Priorities	Process technology	Product technology
Business Priorities	Low cost, high quality	High performance
Production rates	High rates and volumes	Low rates and volumes
Linkage of R&D with Production	Integrated R&D, manufacturing, and service	R&D separately contracted and managed
Technology sharing	Success based on proprietary advantage	Government may require sharing with second source

and correspondingly high political value to both geographic and industrial constituencies. The Apollo space program was surely the grandest of all. Current and recent projects include the Space Station, the Space Shuttle, the National Aerospace Plane, the Superconducting Supercollider (SSC), the Earth Observing Satellite System, and hardware prototypes for the Strategic Defense Initiative (SDI). Funded projects whose cost may run to tens of billions of dollars have great political staying power, even in the face of evidence that the program is unlikely to meet its objectives.[31] For this reason even "small science" projects that are conducted by small groups in widely dispersed scientific laboratories have been bundled by their advocates into administrative megaprojects, such as the Human Genome Project.[32] The financial magnitude of megaprojects can crowd out the more widely decentralized research of universities and private industry. Recognizing this fact, the Clinton administration has announced its intention to redesign the Space Station as a more modest endeavor and to stretch out the SSC to reduce its annual cost in the near term.

The huge national laboratories of the Department of Energy and NASA also represent aggregations of expenditure in specific communities, and generate an equally strong political following. Federal R&D in the middle-sized city of Albuquerque, New Mexico, engages some 6,000 scientists, engineers, and their support staff, working for the Department of Energy and the Department of Defense. As the defense and nuclear weapons programs are drawn down, pressures are rising on these laboratories to diversify, shrink, or find new missions. The direction favored by the administration and Congress is to expand the laboratories' partnerships with commercial industry in the hope of commercializing technologies they have developed. Skeptics of the economic value of this spinoff effort conclude that the laboratories must face up to substantial contraction, such as the defense industry is facing. This issue is the focus of Chapter 4.

New Supply-Side Strategies

There are two primary elements of the new supply-side technology strategy: indirect measures using economic policy and direct investment in R&D. Macroeconomic measures proposed by the Clinton administration would encourage private R&D investments through a permanent R&D tax credit, selective investment-tax credits, modification of capital gains taxation, and other economic incentives. These indirect measures have the advantage of leaving private firms in control of the investment decisions and are relatively easy to administer. They are also relatively expensive. The second supply-side strategy is the investment of federal resources directly in R&D of value to the economy. The federal government invests more than $70 billion in R&D today, most in pursuit of federal missions, as noted above. The administration proposes expansion of four R&D investment programs (each of which is the subject of one of the chapters in this book):

•Funding of "critical" technologies through the Advanced Research Projects Agency in the Department of Defense, and other mechanisms (Chapter 2);

•Funding of advanced technology through cost-sharing agreements between the Department of Commerce and private commercial firms (Chapter 3);

•Funding of research and technology development of commercial interest in the national laboratories (Chapter 4); and

•Funding of research in the universities intended to promote collaboration with industry (Chapter 7).

In addition, last year Congress passed and President Bush signed the Small Business R&D Enhancement Act of 1992, which over the next five years will double the percentage of contract R&D the federal agencies must invest in small businesses.[33] This does not add R&D funds to the federal budget, but will by 1997 shift over $1 billion to small businesses, which are encouraged to commercialize the R&D.

How are federal R&D programs going to enhance the civil industrial base and be flexibly responsive to industrial needs, yet avoid upsetting the delicate balance of market-driven competition, and ensure sufficient distribution of the economic benefits throughout the society so that R&D subsidies to specific firms will be accepted as politically legitimate? Policies governing resource allocation among R&D investments in commercial technologies must address these questions. In recent years a great deal of political attention has focused on Congressional requests that executive agencies prepare lists of "critical technologies," with the implication that technologies so selected would be targeted for preferential support. The emphasis on making lists of "critical technologies" is probably a transient event in U.S. technology policy, although the term is surely here to stay. If new policies are to be implemented that require the selection of specific technologies to be federally subsidized, the criteria for their selection and evaluation must be clearly drawn and related to stated goals. Unless those criteria are robust and widely acceptable, the selection of "critical technologies" will still be viewed as "picking winners and losers" under a different name. This is the subject of Chapter 2.

Even the most robust criteria for selection of technologies are not sufficient to justify government subsidies unless policies are in place to differentiate the government's role from that of private investors and entrepreneurs. The idea that programs meriting priority will be called "critical" will sooner or later have to give way to the reality that the government's role is not made legitimate by its economic "criticality," but by the under-investment by private firms in knowledge that has large positive externalities. Government funding of basic research in low-temperature physics, for example, is surely justified, especially with the prospect that a new industry might arise from the discovery that superconductivity (conduction of electricity without electrical resistance) is possible at much higher and more practical temperatures. Once the attractiveness of ultimate applications puts technologists in hot pursuit a pathbreaking technology can be said to be in prospect. But just putting superconductivity studies on a list of "critical" technologies fails to justify federal support for such research. That justification must come from recognition that basic research is a public good and that individual firms can be expected to invest less in pathbreaking technology than would be justified by estimates of potential social return. The economic criteria for defining the federal role in commercial R&D are the subject of Chapter 3.

Signs that the policy debate is becoming somewhat more sophisticated are at hand. The National Competitiveness Act of 1993[34] has a Title III called "Critical Technologies," but rather than calling for another list of critical technologies, it authorizes a formal process of technology "benchmarking." Benchmarking is the analysis of the competitiveness of the scientific and technological capabilities of American firms relative to those of other nations. The bill would give the Department of Commerce lead-agency responsibility for establishing a process for such analyses. The bill does not prescribe how this net technical assessment information is to be used in allocating resources of commercially relevant R&D, but the need for competitive and comparative assessment of United States industrial technology is clear. One may hope that the work on priority-setting for federal R&D, already initiated by the Subcommittee on Research of the House Committee on Science, Space, and Technology, will result in the development of clearer, more appropriate language in this area.

Demand-Oriented Strategy: Information and Collaboration

The application of technical knowledge to every phase of the innovation process is accelerating. New approaches to innovation permit a drastic reduction in time between the recognition of a market need and its satisfaction in volume production, thus enhancing consumer satisfaction and dramatically reducing the cost of product and market innovations. Technology also diffuses much more rapidly as technical knowledge becomes codified, dramatically increasing the speed of technology transfer compared with the transfer of tacit, or embedded, knowledge.[35]

Furthermore, with accelerating mobility of capital, technology, and people, and the steady erosion of protectionist barriers to trade, national boundaries are rapidly losing their significance as natural domains within which innovations are generated. Transnational firms can more easily initiate production in many geographic locations simultaneously. One consequence is that any government's ability to control technological outcomes unilaterally within its borders is weakened. Rates of technology diffusion within a national economy must, therefore, be even faster than in competing nations if domestic establishments are to have a competitive advantage from government investments in public knowledge. The acceleration of technological diffusion thus becomes a major element of national competitiveness strategy.

Present policy has given the United States a very strong research capability. If as much attention were given to finding, acquiring, adapting, and using technology as to creating it, Americans could take better advantage of the nation's leadership in research. Otherwise, the case for investing more resources in new technology will be undermined by evidence that foreign firms are better positioned to profit from it than American companies.

Federal policy should be shifting toward activities that are better designed to help firms acquire and use available technology. Such activities could make a more immediate contribution to technological performance. More emphasis should be placed on education, skills enhancement, and technology diffusion and absorption. What America needs is a "capability-enhancing" technology policy. The facilities and services that address these needs must be appropriate for all non-proprietary technology, not just government-generated technology. The scope of federal agency missions is a poor surrogate for the technological interests of commercial manufacturing.

Americans must also overcome their habit of ignoring technologies that were "not invented here." In this respect, the business community is ahead of the politicians, who focus their attention on the inaccessibility of Japanese markets, rather than on learning from Japanese successes. Even if the trade "playing field" were level, U.S. firms would still have problems competing with Japan in microelectronics, computers, machine tools, opto-electronics, and communications equipment. Trade policy is a poor surrogate for technology policy.

Government Encouragement of Inter-firm Collaboration

A key element of any diffusion strategy will be the encouragement of "pre-competitive" collaboration among like-minded firms, both with one another and with universities and government laboratories. A quiet revolution has taken place in the government's attitude toward anti-trust constraints on strategic alliances and cooperative research, and the business community is becoming more comfortable with the new attitude. This policy has been strongly influenced by the perception that inter-firm cooperation has been a competitive advantage in Japan, and perhaps in the European Community as well. As American industries have lost world market shares, it is easier for government to view consortia and other forms of inter-firm cooperation as legitimate responses to competitive pressures from abroad.

The high mobility of codified scientific and technological knowledge and skill leads governments to question whether investments in research in the public domain reserve enough of the benefits for their own citizens, rather than those in other nations. Those governments with good working relationships with their industries may be able to capture more of the benefits from research locally, while exerting influence far beyond their borders. The funding of pre-competitive R&D in the Advanced Technology Program, whose costs are shared by industry, is an example of an attempt to do this. There are three reasons why pre-competitive research collaboration may be appropriate and effective technology policy:

•Joint ventures in pre-competitive R&D may permit more rapid response to changes in markets or technology than will arise from fragmented or monopolistic markets.

•Collaboration in pre-competitive R&D can accelerate the diffusion of technology, and, most important, of the tacit knowledge that is acquired only through shared experience.

•Shared R&D of a pre-competitive nature can contribute to the mutual confidence required for business joint ventures, thus reducing political barriers and lowering the transaction costs of industrial alliances.[36]

The high level of interest in Europe, in Japan, and more recently in the United States for government policies to encourage government-sponsored, R&D-based industrial collaboration may seem surprising in view of the traditional suspicion of government interference on the part of private company executives. In Japan, for example, many of the largest firms, which are fiercely competitive, are quite ambivalent about pressures from the Ministry of International Trade and Industry (MITI) to engage in government-sponsored industrial cooperation with one another. In addition, multinational firms may have difficulty in acceding to the wishes of a host government, if those wishes are contrary to the interests of another host government where they also operate. However, in each of the three regions there are excellent reasons why government-fostered R&D collaboration have nevertheless appeared attractive.

In Europe, as noted above, the challenge was—and is—market integration. The first attempt to build a Europe-wide base in the computer industry—UNIDATA—took that challenge head on. It was, however, a failure—imposed on national champion firms by governments that had not first built the interrelationships and trust that cooperative R&D might have facilitated.[37] The European Strategic Program for Research and Development in Information Technology (ESPRIT) took the opposite approach and subsidized pre-competitive, collaborative research among European computer firms (including IBM subsidiaries), and has been more successful.

Japan's interest in R&D collaboration derives from the nation's keen sense of vulnerability to changes in the outside world, a sense greatly exacerbated by the oil shock of 1973. While Japanese high-tech firms remain highly competitive in domestic markets, they are much more willing to collaborate with one another to gain shares in major export markets. Studies of Japanese research collabora-tion conclude that the leadership role of MITI as the originator of such national efforts is greatly exaggerated by foreign observers. However, once a consensus in the business community to do so has been reached, MITI becomes the organizer and the enforcer of a national effort to increase market shares for Japanese firms in world markets. The Japanese government funds a smaller percent-age of R&D (only about twenty-one percent) than do governments in North America and Western Europe. A strategy based on the commercialization of government-generated R&D, such as that implied by the proposed missions in technology transfer for United States national laboratories, would have much less effect in Japan than in the U.S., where government funds a much larger share of national R&D.

The United States has only recently started to encourage sharing of R&D by consortia of companies, both because of traditional political commitments to *laissez-faire* economic policy and because of a traditional reliance on anti-trust law as a means for fostering competition. The formation of the Microelectronics and Com-puter Corporation (MCC)—a consortium of fifteen United States-based computer companies, but not including IBM or ATT—led to Congressional passage of the National Cooperative Research Act of 1984.[38] Firms that register with the Department of Justice as a consortium for pre-competitive R&D are shielded from treble damages in the event they are sued for civil anti-trust violations. By implication, they are also less likely to be prosecuted by the government. The number of consortia registered under this law has been rising rapidly.

The Clinton administration proposes to accelerate this trend toward cooperative industrial research by extending the National Cooperative Research Act to authorize cooperation in manufactur-ing as well as research. A specific application of "information infrastructure" is also aimed at this objective: the administration's plan proposes that "agile manufacturing programs [should be] expanded to allow temporary networks of complementary firms to come together quickly to exploit fast-changing market opportunities."[39]

The current pattern of industrial technology assistance in the United States is more fragmented and less developed than in Japan and in several European countries.[40] The 1988 Omnibus Trade and Competitiveness Act charged the Commerce Department's National Institute for Standards and Technology (the former National Bureau of Standards) to help the states establish Manufacturing Technology Centers (MTCs) to work with small to medium-size manufacturers to improve their productivity. The Clinton administration also proposes a major expansion of these industrial extension services to small and medium-size businesses. To be successful, this effort will require the cooperation of both the business community and the states. Middle-size or small firms frequently lack the expertise to adopt new technologies that might enhance their productivity. An effective extension program would strengthen manufacturing capabilities, provide high-quality, cost-effective inputs to other manufacturers, and contribute to reducing the trade deficit. Manufacturing extension is the subject of Chapter 6.

A pressing need is for an efficient technological infrastructure, especially for the diffusion of information, explored in Chapter 5. This is one of the stated priorities in Allan Bromley's *United States Technology Policy*, and is also a high priority of the Clinton-Gore administration. The signing in December 1991 of the High Performance Computing Act authorizes the upgrade and rationalization of the federally supported portions of the Internet into a "National Research and Education Network" (NREN), serving schools, universities, government, and industry.[41]

The NREN is conceived as a program engaging the efforts of both public and private enterprise. It has the potential to create a national market for both commercial and not-for-profit information services that could make a big contribution to technological productivity. It may greatly expand the capacity for collaboration by geographically dispersed firms and institutions. Thus it lends itself to the efficient operation of consortia and other collaborations through which public knowledge can be effectively shared. The high priority the Clinton administration attaches to this activity derives both from the president's conviction that infrastructure investments are important to the economy, and to Vice President Gore's personal association with the High Performance Computing Act, which is popularly called the "Gore bill."

From a long-term perspective the highest priority is the need for more attention to research, education, and training to enhance the capabilities of the workforce and to build professional technical competence, especially in the "downstream" technologies: production processes and manufacturing systems. The Clinton-Gore policy paper proposes to expand the ability of technical and community colleges to attract students to careers in manufacturing engineering. Universities are collaborating with industrial engineers in a number of National Science Foundation funded Engineering Research Centers, many of which are devoted to manufacturing processes. Chapter 7 examines the position of the research universities, which are increasingly called on to make more direct contributions to industrial competitiveness through both training and research. Chapter 8 examines the human resource development component of the technology strategy. It suggests that an elite but limited cadre of technical professionals will not be successful at restoring American productivity growth rates without support from a better-trained workforce, and calls for a culture change in which skilled work by the non-college trained is more highly respected and fostered. It notes that the administration's proposals for supporting the college-bound through a Service Corps might prove a disincentive for this culture change. No technology strategy can overcome the productivity drag of an underskilled, demoralized, or undermotivated workforce.

Challenges to the Successful Implementation of New Technology Policies

There are five primary obstacles to consensus on technology policy as it relates to the civil economy:

•Lack of crisp, robust criteria to delimit the federal role in each of the program types devoted to commercial technology;

•Concern that even if such criteria are established, they will be corrupted politically;

•Lack of public confidence in the government's ability to build institutions with the competence and the latitude to make technical and economic judgements—a view that questions agencies' legitimacy;

•Confusion about policies toward direct foreign investment and how to deal with U.S. firms that are subsidiaries of foreign corporations, and with the overseas subsidiaries of U.S.-headquartered firms; and

•The fact that decentralized programs of technical investment lack a strong constituency and thus have difficulty competing with the focused constituencies that coalesce around the concentrated spending of megaprojects.

One obstacle to consensus on technology policy is expressed in the question most often posed by economists: "What proof," they ask, "have you that U.S. industry is in technological decline, requiring government intervention?" This is coupled with a question asked by conservatives: "What proof have you that the government has the competence to help, even if such help were indicated?"

To the first question one might reply, "It does not matter." In fact, there are many signs of resurgence of the United States' manufacturing competitiveness. Even the merchant semiconductor industry has reversed its trend toward lower world market share, and is showing a small gain. Even if the economy were gaining strength and employment were high, the case can be made that the U.S. government should continue to enhance the competitive advantage of its institutions and people. Indeed, firms are more likely to make use of government-sponsored R&D and related capability-enhancing investments if the industry is more prosperous and is investing for the future. If it is cutting costs, reducing R&D, and laying off people, the availability of new technology and new tools for using it may be of secondary interest.

As for questions about government agency competence: strategic technological assistance to save a moribund segment of industry usually faces a poor prognosis. But the U.S. government has demonstrated for forty years the ability to create basic science for the world and technology for its own missions. Although megaprojects are a counter-example, agencies nonetheless have substantial experience with the very kind of decentralized, merit-based, competitive programs that are needed for supporting pathbreaking and infrastructural technologies. Government agencies must, however, develop robust, disciplined policies, staff the work with talented people, and solicit the cooperation of technical experts in industry to help manage the new programs.

There remains this question: What constituency will support these emerging technology policies and shield the politicians from the heat of criticism when conflicts arise, as they surely will?

Only the business community, supported by labor and the technical community, can provide legitimacy and effectiveness to technology policy. Business leaders must articulate the economic and political philosophy that allows a role for government in the encouragement of pathbreaking and infrastructural technologies. If they do, political leaders will follow their lead. Here one must be impressed by the extent of the private sector's support for a well-constructed federal program of investment in technology infrastructure. The National Association of Manufacturers, the Council on Competitiveness, the Computer Systems Policy Project, and trade groups such as the American Electronics Association and the Aerospace Industries Association are consistent in their call for a more determined and sustained effort to enhance the technical comparative advantage of U.S. firms.

How well is the U.S. government organized to carry out such a policy? The structure of its technical agencies resulted from a long period of optimizing government technology investments around defense and other federal big-science missions, a *laissez-faire* approach to commerce, and a commitment to academic autonomy in the management of federally-funded basic research. It functioned well in these roles but does not seem well adapted to the new policy environment, which calls for decentralized, collaborative efforts with industry. State governments compete with each other in fragmented efforts, with funding often too unstable for long-term efforts. Each state negotiates separately with each federal agency.

Every federal agency that is likely to play a major role in the attempt to enhance industrial competitiveness has some drawback impeding that role. The Commerce Department is relatively inexperienced with large-scale technology management, and receives traditionally weak support from its business constituency. The Department's National Institute for Standards and Technology (NIST) is industrially experienced, but it is a laboratory, not structured as a primary operating agency for civil technology policy. The Defense Department is ill-suited to supporting commercial innovation because of its command economy culture. Acquisition reform is prerequisite to easy defense access to commercial industry. In any case, rapidly declining budgets will narrow the focus on military priorities. The Department of Energy has very extensive R&D resources in its national laboratories, but its nuclear weapons operations are traditionally vertically integrated and internally self-sufficient, giving them little commercial experience. NASA has a relatively narrow mission, and needs to focus all its energies on clarifying its future mission and improving its efficiency and performance. The National Science Foundation traditionally focuses on academic basic research; its expanded activities in engineering and in collaborations with industry are resisted by disciplinary scientists, who see these activities as being undertaken at the expense of basic research. The National Institutes of Health is focused on basic and clinical biomedical research. Its success in stimulating biotechnology industry is creating tensions between its basic science mission and the Congressional mandate to collaborate with industry. In Chapter 9 we will present ideas for overcoming these shortcomings and will suggest how their roles may evolve.

Changing the structure of agencies in the executive branch is difficult under any circumstances. It is made even more difficult when the administration's list of legislative action it expects from Congress is dauntingly large, and when the structure of the Congressional committees responsible for the agencies is itself a barrier to needed changes in that structure. Thus minimal organizational changes should be expected. The resulting persistence of old patterns of agency and constituency behavior will be an additional challenge to implementing the new policies.

Several challenges for the management of a new technology policy flow through the following chapters:

•Even as nations experiment with new paradigms for industrial policy, the nature of the innovation process in the most successful firms is changing rapidly. Public policy must adapt to a moving target even as it is being formed and implemented.

•Public policies for enhancing the technological dimensions of competitiveness, and the assumptions underlying these policies, must reflect differences, often quite striking, among the industries concerned. Thus macro-level policies are unlikely to be successful unless interpreted quite flexibly when applied to different industries.

•U.S. government policies cannot be developed independent of comparable activities in U.S. trading partners. Although all industrialized nations face similar challenges and opportunities, their policies may be quite different; technologies flow quickly around the world, and the business context in which they are developed and used is global.

•The role of government grows more, not less, complex: with the increasing trend toward private initiative and private resources deployed in cooperation with government, the government must share decision-making with its private partners. The economic health of the world will, paradoxically, depend even more on wise government policies, as government action becomes more limited in scope and more restrained in nature.

•Since many of the new federal technology initiatives imply the collaboration of states with federal agencies, the states will have to seek an institutional mechanism for speaking with a more unified voice when working with federal agencies.[42] The new federal administration has not yet confronted this issue, although the president's experience as governor of Arkansas equips him to do so.

•Initially, the Clinton administration priority on domestic economic concerns will have the effect of drawing attention away from the international dimensions of science and technology policy. Given the strong international interdependence of science and technology, and the important place that foreign trade plays in the world economy, international cooperation in science and engineering will need to be given a higher priority.

•There will be serious political hazards from what is unavoidably a "trickle down" policy that converts public assets to private ones in the quest for more jobs and more wealth. The politics of redistribution is both impatient and suspicious. The political hazards must be mitigated by limiting government investments to activities that compensate for demonstrable market failures.[43] These pressures may also lead to pressure for a more explicitly geographical redistributive policy for program selection and management, although this has not yet appeared.

The Clinton-Gore strategy is clearly rooted in confidence in the efficacy of government action to promote innovation. They declared that:

The traditional federal role in technology development has been limited to support of basic science and mission-oriented research in the Defense Department, NASA, and other agencies. This strategy was appropriate for a previous generation but not for today's profound challenges. We cannot rely on the serendipitous application of defense technology to the private sector. We must aim directly at these new challenges and focus our efforts on the new opportunities before us, recognizing that government can play a key role helping private firms develop and profit from innovation.[44]

The Bush and Reagan administrations shared—indeed fostered—the skepticism of many Americans that the Clinton administration's confidence in a vigorous technology policy is justified. Whether this confidence and the skill of the new administration to define and implement the new policies can overcome the opposition of ideological conservatives and the skepticism of the voting public remains to be seen. But by the same token, a reasonable measure of success could not only reward the nation with a more vibrant economy and the rejuvenation of its technical capabilities, but can go a long way toward allaying that skepticism.

Notes

1. See Lewis M. Branscomb, "Technology Policy and Economic Competitiveness," *Science and Technology Policy Yearbook 1992* (Washington, D.C.: American Association for the Advancement of Science, 1993).

2. John A. Alic, Lewis M. Branscomb, Harvey Brooks, Ashton B. Carter, and Gerald L. Epstein, *Beyond Spinoff: Military and Commercial Technologies in a Changing World* (Boston: Harvard Business School Press, 1992), chap. 1.

3. William J. Clinton and Albert Gore, Jr., *Technology for America's Economic Growth: A New Direction to Build Economic Strength* (Washington, D.C.: The White House, February 22, 1993). See also Clinton-Gore Campaign, *Technology: The Engine of Economic Growth: A National Technology Policy for America* (Little Rock, Ark.: September 21, 1992).

4. The elements of the Clinton administration policy include: (a) support for R&D in support of the commercial industrial base; (b) an aggressive program to build 172 manufacturing extension centers, building on the Manufacturing Technology Centers established by the Department of Commerce; (c) investment in information infrastructure, including extending the National Research and Education Network program to schools, libraries and hospitals; and (d) a focus on human resource development—with one welcome addition to current federal activities: an apprenticeship program on the European model to address school-to-work transition for the non-college bound.

5. Pre-competitive R&D refers to research conducted prior to the creation of manufacturable prototypes ready for market introduction. Infrastructural research is research broadly applicable to industrial development, motivated by technological problems rather than intrinsic scientific interest. See Chapter 3 for more complete discussion of both.

6. U.S. Code 6683.

7. Nixon was able to do this by acting just before the January 31, 1973, expiration of the legislative authority to reassign existing statutory functions among the agencies. President Kennedy had used that same reorganization authority to form OST in the President's Executive Office out of the pre-existing Federal Council for Science and Technology, which had been established by President Eisenhower.

8. Ibid., Section 6602 (b) (3).

9. Ibid., Section 6602 (c) (2).

10. Allan Bromley, *The U.S. Technology Policy* (Washington, D.C.: The Executive Office of the President, September 26, 1990).

11. P.L. 100-418.

12. See Branscomb, "Technology Policy and Economic Competitiveness."

13. See L.M. Branscomb, "Information Infrastructure: A Public Policy Perspective," in Brian Kahin, ed., *Building Information Infrastructure: Issues in the Development of the National Research and Education Network* (New York: McGraw-Hill, December, 1991).

14. Clinton-Gore campaign, *Technology: The Engine of Economic Growth*, September 21, 1992.

15. Clinton and Gore, *Technology for America's Economic Growth*.

16. The title "Science and Technology Advisor" is the commonly used informal title. Dr. Gibbons' formal White House title is Assistant to the President for Science and Technology.

33

The National Technology Policy Debate

17. For a historical analysis of postwar U.S. science policy, see Bruce Smith, *American Science Policy Since World War II* (Washington, D.C.: The Brookings Institution, April 1990). For an analysis of U.S. technology policy, with emphasis on national security and economic concerns, see Alic, et al., *Beyond Spinoff*. For an annual summary of issues, documents, and data about U.S. science and technology policy, see Margaret O. Meredith, Stephen D. Nelson, and Albert H. Teich, eds., *Science and Technology Policy Yearbook 1991* (and annually) (Washington, D.C.: American Association for the Advancement of Science, 1991 and annually). Other recommended references will be found in the Bibliography at the end of this book.

18. Vannevar Bush, *Science—The Endless Frontier*, A Report to the President on a Program for Postwar Scientific Research (Washington, D.C.: Office of Scientific Research and Development, July 1945; repr. National Science Foundation, 1980), pp. 5–40.

19. Bruce L.R. Smith, *American Science Policy Since World War II* (Washington, D.C.: Brookings Institution, 1990), pp. 48–52, and 164–166.

20. Henry Ergas, "Does Technology Policy Matter?" in Bruce R. Guile and Harvey Brooks, *Technology and Global Industry: Companies and Nations in the World Economy* (Washington, D.C.: National Academy Press, 1987), p. 192.

21. Alic, et al., *Beyond Spinoff*, pp. 88–99.

22. In 1960, government expenditures on R&D dominated private investments ($8.7 billion to $4.5 billion). In that year 80 percent of federal R&D investments were from defense. In 1963, if atomic energy and space are included with defense, they comprised 93 percent of the federally supported R&D effort and more than two-thirds of the entire national effort, public and private. In effect, defense *was* the national effort.

23. This is less true, however, for smaller firms that supply both defense and civil markets. Manufacturing establishments using metalworking tools account for twenty to twenty-five percent of manufacturing employment in both defense and civil sectors. These are in effect dual-use firms. See Maryellen R. Kelley and Todd A. Watkins, "The Defense Industrial Network: A Legacy of the Cold War," *Contractor Report* (Washington, D.C.: U.S. Congress, Office of Technology Assessment, November 1992).

24. Stephen Kline, "Models of Innovation and their Policy Consequences," in Hiroshi Inose, M. Kawasaki, and Fumio Kodama, eds., *Science and Technology Policy Research: "What should be done? What can be done?"* (Tokyo: Mita Press, 1991), pp. 125–140. For background, see Stephen Kline and Nathan Rosenberg, "An Overview of Innovation," in Ralph Landau and Nathan Rosenberg, eds., *Positive Sum Strategy: Harnessing Technology for Economic Growth* (Washington, D.C.: National Academy Press, 1986), pp. 275–306.

25. Ralph Gomory, "From the 'Ladder of Science' to the Product Development Cycle," *Harvard Business Review*, November–December 1989, pp. 99–105.

26. Alic, et al., *Beyond Spinoff*, p. 9.

27. President's Scientific Research Board, *Science and Public Policy: Administration for Research*, 3 vols. (Washington, D.C.: U.S. Government Printing Office, 1947), Vol. 1, usually referred to as the "Steelman Report," p. 26. These recommendations were carried out by President Eisenhower in Executive Order 10521, March 17, 1954.

28. Edwin Mansfield, "Academic Research and Industrial Innovation," *Research Policy*, Vol. 20 (1991), pp. 1–12.

29. Clinton-Gore, *Technology for America's Growth*, February 22, 1993, p. 8.

30. "Dual-use technology" refers to a technology whose potential uses include those of both military and civil importance. Technologies for designing, making and using jet engines, optical fiber cable, magnetrons, boron fiber composites, rocket propellants, infrared sensors, ion implanters, computers, 3-D hydrodynamic codes, and automated turret lathes are all "dual-use technologies." The phrase has its origin in export control regulations; a product or a technology may fall under the Munitions Act definitions as a "munition," yet have its primary role in commercial markets with no military connections.

31. Linda Cohen and Roger Noll, *Technology Porkbarrel* (Washington, D.C.: The Brookings Institution, 1991).

32. Mel Horwich, "From Unitary to Distributed Objectives: The Changing Nature of Major Projects," *Technology in Society*, Vol. 12, No. 2 (1990), pp. 173–196.

33. Public Law 102-564. Under this statute, Small Business Innovation Research (SBIR) grants grow from 1.25 percent of agency contract R&D to 2.5 percent by 1997–2000 and the dollar amounts of grants are doubled.

34. H.R. 820, introduced February 4, 1993.

35. Tacit knowledge is that gained through on-the-job experience and is transferred through collaboration or apprenticeship over extended periods of time. Many technicians are highly skilled at specialized tasks but cannot document exactly how they accomplish them; such skill is said to be embedded in the workers' experience. See Alic, et al., *Beyond Spinoff*, pp. 30–34.

36. I am grateful to Todd Watkins of Lehigh University for these observations, which are drawn from an unpublished dissertation for Harvard University's John F. Kennedy School of Government.

37. UNIDATA was a computer marketing corporation created in the 1970s at the urging of the German, French and British governments to market the products of the three main Europe-owned computer companies: Siemens, Honeywell-Bull, and ICL Limited. It was a failure, and spelled the end of European confidence that national champion companies supported by their governments could compete successfully with IBM's European subsidiaries.

38. Public Law 98-462.

39. Ibid., p. 10.

40. Philip Shapira, *Modernizing Manufacturing* (Washington, D.C.: Economic Policy Institute, 1990).

41. Public Law 102-194.

42. Richard Celeste, former governor of Ohio, chaired a study for the Carnegie Commission on Science Technology and Government, which addressed this issue: *Science Technology and the States in America's Third Century* (New York: The Carnegie Commission, September 1992).

43. See Chapters 3 and 4 of this book; and Alic, et al., *Beyond Spinoff,* chap. 12.

44. Clinton-Gore, *Technology for America's Economic Growth,* p. 1.

2

Targeting Critical Technologies[1]

Lewis M. Branscomb

The term "critical technology" carries a certain amount of ambiguity. How do you decide what is critical? Critical for whom?.... Finding a critical technology is not the problem. The problem is what to do with it once you've found it.

Robert White, Undersecretary of Commerce for Technology, NIST, Gaithersburg, Md., February 6, 1992.

During the Bush presidency the Democratic majority in Congress began to apply pressure on the administration to give a higher priority in its national technology policy to economic goals. One focus of this pressure was to urge the administration to invest a larger share of federal R&D expenditures in the development of technologies useful to commercial firms, particularly those in industries hard pressed by foreign competition.[2] The Congress attached to various appropriations bills requirements that agencies make lists of "critical technologies" that might deserve special government attention (whether for protection or for promotion).

This focus on specific technologies of commercial interest represents a major departure from the traditional criteria: in the past, technologies selected for federal investment were those required for the performance of federal agency missions, such as weapons systems, space technology, or air traffic control equipment. The new lists were meant to identify technologies thought to be particularly critical to increased industrial competitiveness. In this way technologies would be placed on the federal R&D agenda that had not received priority attention in the past, when almost all federal technology development was in pursuit of specific mission objectives. Now, agencies would undertake research related to production processes, design tools, industrially useful materials, and other capabilities needed by industrial firms competing in commercial or government markets, or both.

As a guide for allocating government R&D resources, however, the making of lists of critical technologies leaves many questions unanswered. Policies aimed at identifying and generating "critical technologies" may be necessary, but they are by no means sufficient to achieve economic growth, a cost-effective military, a cleaner environment, or any other social goal where global industrial competition determines outcomes.

Much of the public discourse about the importance of critical technologies begins and ends with a plea for excellence in new technology generation, but the value to society of technical excellence depends on its successful application. A leadership position in a technology, however exciting and important, does not, of itself, assure prosperity, a strong defense, or a clean environment.[3] First, technologies must be mastered, reduced to practice, supported by cost-effective production processes, and introduced to the market. Then that market position must be sustained by appropriate complementary assets, by effective channels of distribution, and by responsive customer service.[4] Even that, however, is not enough, for many innovating products have found strong initial markets, only to see other firms—sometimes in other nations—capture the lion's share of market growth through incremental functional improvements, cost reductions, quality superiority, and better marketing and service. For example, the British firm EMI, Ltd., invented the medical "CAT-scan" X-Ray device, but within a very short time much of the market had been captured by GE in the United States.[5] IBM had remarkable success with the personal computer, capturing at one time 45 percent of the desktop computer market, only to see it erode to manufacturers of lower-cost, compatible "clones."

As Harvey Brooks observes:

[The] current United States mental model of innovation places excessive emphasis on originality, in the sense of newness to the universe as opposed to newness in context....In other words, our present ways of collecting statistics about science and technology are focused primarily on R&D, often leading policy-makers to equate innovation policy with R&D policy. It leads them as well to overemphasize the significance of originality—of "newness to the universe"—as the sole source of the economic and social benefits arising from R&D or scientific and technological activity more broadly.[6]

Richard Nelson defines innovation as "the processes by which firms master and get into practice product designs and manufacturing processes that are new to them, whether or not they are new to the universe, or even to the nation."[7] Perhaps a simpler definition would be "a process leading to creation and successful introduction into the market of a product or service new to the firm." As we shall see, the "critical technologies" paradigm behind the making of critical technologies lists fails to address the circumstances and processes necessary for a technology to be incorporated in a successful innovation.

In one respect, however, the introduction of "critical technologies" into the vocabulary of debates about industrial policy represented a subtle but important step forward. The *laissez-faire* policies of the Reagan and Bush administrations supported the federal funding of basic research as a necessary "public good" investment, but took the view that "downstream" elements of the system of industrial innovation (design tools, manufacturing engineering, product testing and the like) required specific market knowledge and must therefore be left to the private sector to support.[8] Now that the word "technology" has been introduced into the dialogue about federal support for non-military R&D, all those "downstream" engineering activities become part of the package. The political objection, often made during the Reagan presidency, that federal involvement in commercial downstream engineering activities constituted an unacceptable form of industrial policy, was thereby finessed.

The primary issues addressed in this chapter are:

•Although a government role targeting industrial technologies is well established in Europe and Japan, this role is far less established in the United States. Here it has mostly focused on the export control regulations in the 1970s. This history suggests that the U.S. government's new interest in critical technologies has both promotional and restrictive aspects that may conflict. The more recent history suggests that Congressional pressure for lists of candidate technologies for government support represents an effort to get civilian industrial policy higher on the nation's economic agenda.

•The federal government faces a number of problems when it relies solely on the critical technologies paradigm to allocate resources among competing claimants. These problems include the association of listed technologies with "crystallized interests" of benefiting firms, the arbitrariness of criteria for selection, the high level of abstraction of identified technologies, and the temptations of protectionist constraints on international exchanges of technical knowledge. Perhaps the most serious deficiency is the need to distinguish the government's role in the listed technologies from the role of the private sector.

•After the technologies are identified, what policy choices remain to be made? How can these choices be integrated into an effective policy for enhancing industrial performance? What forms of institution are most suitable for managing technology selection? In the United States these vary widely, from government support for industry consortia, to proposals for a "Civilian Technology Corporation," or a "Civilian DARPA"—an agency which presumably would grant R&D contracts for technologies identified as critical to some economic objective.

•What is the future for the critical technologies concept? Perhaps the best explanation for the emergence of critical technology studies is the recognition that the focus on missions that has defined technology priorities in the past must yield to a new way to set priorities in support of economic health. For government to be an effective partner, it should undertake systematic assessments of the competitive condition of U.S. industrial technology compared to the "best of breed" abroad. But "critical technologies"—lacking well-defined criteria for choosing the technologies, and lacking associated policies for commercialization and diffusion of the technology—is not up to the task.

We conclude that "criticality" is a poor way of identifying commercially relevant R&D projects that merit public funding. It is too context dependent and too vulnerable to political manipulation to be a useful criterion for choice. The legitimacy of the government's role in commercial technology is not defined by perceptions of abstract importance, but by the nature of the government's relationship to the private sector, which is responsible for translating those technologies into jobs, wealth, and trade. Nevertheless priority choices for federal R&D goals must be made. We conclude that government priorities should be set by a combination of criteria rooted in economic theory, which circumscribe the appropriate role of government in the commercial economy and are linked to commitments from the private sector to contribute to and make effective use of the technology. We recommend that words other than "critical technologies" be used to describe priority areas of science and engineering, to avoid not only its ambiguity, but also its history of association with government control that may encourage protectionism.

Origin of "Critical Technologies" in U.S. Technology Policy

It was in the United States that the selection of lists of "critical technologies" as a preliminary step in formulating industrial policy first came into vogue. Of course most nations have indulged in industrial policy, as noted above. Insofar as they supported commercial technologies on the basis of a set of priorities, those selected might be considered more "critical" to national well-being than others.

Critical Technologies for Purposes of Export Control

In the United States, the focus on critical technologies has its roots in defense technology policy. On February 27, 1976, the Defense Science Board (DSB) published a task force report, commonly referred to as the "Bucy Report" after the chairman, Fred Bucy of Texas Instruments Co., which recommended a radical reform of export control strategy.[9] It proposed that controls should be focused on retarding transfers of technology that could significantly enhance the military capability of potential adversaries. This strategy was seen as particularly useful when applied to turn-key facilities with which both weapons and civilian products could be produced. Thus the new policy would address a troublesome area not adequately dealt with by export controls—commercial technology with both civil and military capabilities. Such technology is called dual-use technology.[10] Indeed, one benefit to industry was expected to be reduced reliance on the Commerce Department's Controlled Commodities List (CCL), a formal list of products of military importance whose export was forbidden without a government license. Bucy hoped that if controls were created for dual-use technology, the Commerce Department would drastically reduce the number of products restricted by the CCL, thus reducing barriers to U.S. exports. The justification for this liberalization of controlled commodities was the view that reverse engineering of manufactured products (the art of deducing how a product was made from an analysis of its components) would prevent hostile nations from gaining production capabilities by purchasing American products. The DSB recommendations were embodied in the Export Administration Act of 1979,[11] as amended in 1985, which emphasized the control of specified critical technologies and assigned this responsibility to the Department of Defense.

A key step was the creation by the Secretary of Defense of a "list of militarily critical technologies," which was published in classified form in October 1986. This Militarily Critical Technologies List (MCTL) "provides a detailed and structured technical statement of development, production, and utilization technologies which the Department of Defense assesses as being crucial to given military capabilities and of significant value to potential adversaries."[12] It covers private as well as government-owned technologies, and is used by government officials to make judgments about licensing exports of the know-how described in its pages. These judgments are necessarily highly subjective, and have been the source of much controversy.

The statute says that primary emphasis shall be given, *inter alia*, to arrays of design and manufacturing know-how, keystone manufacturing, inspection, and test equipment, and goods accompanied by know-how or which might give insight into the design and manufacture of a U.S. military system. Thus the MCTL is not a list of critical military material and weapons systems. It is instead a list of technological capabilities which, if acquired by firms in a potentially hostile state, would give that state the capability to erode a military advantage the United States perceives to be essential.

If the industrial capability in question has "dual uses," a factory possessing that capability can, in principle at least, produce civilian goods as well as weapons systems. (The Controlled Commodities List did not have this property, for it refers to specific single products.) A "militarily critical technology" is likely to include a family of tools and know-how which only in combination create the production capability in question. Increasingly, technologies of military importance are driven by huge commercial markets. For example, in the last decade of the Cold War, as both nuclear weapons and their delivery systems were approaching saturation and equality on both sides, strategic advantage began to depend more on command, control, communications, and intelligence. The primary technologies for these functions are derived from the electronics, communications, and computer industries, where commercial technology leads military technology by substantial margins. Thus these areas of military importance increasingly depend on dual-use technologies.

Two consequences flow from this observation. First, the criteria for "criticality" for export control purposes begin to resemble those one might apply in a policy designed to enhance commercial competitiveness. Consider a dual-use technology identified with "critical" military interests. A firm seeking commercial export markets for the technology's non-military applications might resist the imposition of export controls based on that label. But to a defense firm, designating a technology as "controlled" might mean that it is a defense investment priority; to such a firm, a "militarily critical" label might be a favorable attribute of a technology, suggesting its importance to a government customer.

Second, because the criteria for "criticality" for export control purposes and those for priority investment began to converge, some trading partners of the United States became concerned, even as they cooperated with the United States in the NATO Coordinating Committee (COCOM) for export control, that the MCTL might become an instrument of trade protection under the guise of a tool for protecting collective security.

Thus the origin of critical technologies in export control terminology carries an implication that both security and economic interests are involved. This linking of military concerns with the desire to support industrial competitiveness suggests the danger that protectionist policies might become attached to those technologies deemed "critical" and receiving government subsidy.

Identifying Technologies for Commercial Promotion

That governments might select specific industries for preferential treatment, and pursue that preference through investments in the core technologies for the industry, is not a new idea. After World War II, governments in both Europe and Japan invested directly to rebuild the scientific and technical underpinnings of their industries. In several countries, key industries were nationalized or supported by major injections of public capital. Subsidies for the associated technology supported these capital investments. After an unsuccessful attempt to integrate the European market for "national champion" computer companies, the European Community began in the 1970s to encourage cross-national collaboration in pre-competitive R&D (that is, R&D that precedes the development of a commercial product in a form ready for production).[13] In this effort to help national firms compete more successfully with North American and Japanese multinationals, certain technologies were identified as of particular importance. Technologies on which the computer industry is based formed the European Community's ESPRIT program, the first of a series of research and development efforts funded by the European Community (EC) and by participating firms from more than one European country, and referred to as "framework" programs. Telecommunications, biotechnology, and other framework programs were added in time. However, the selection of specific technological goals has generally been left to the firms and research institutions proposing to work within the framework and through another EC program called Eureka. Strategic selections are usually made at the industry level.

The same is true of most of the Japanese programs coordinated by the Ministry of International Trade and Industry (MITI). MITI programs were launched in support of national priorities to be given to specific industries—for example, automobiles, ship building, and aircraft. However, MITI also launched a number of technology programs, of which the best known are the Very Large Scale Integration, Fifth Generation, and Super-Computer programs. These endeavors had specific technical goals, and might be considered national programs to develop "critical technologies," although they were not characterized as such.

In the late 1960s, under the Kennedy administration, the Department of Commerce launched an initiative to support selected "civilian industrial technologies." Lagging industries, where unemployment was rife, were selected as targets for technical help. The administration launched a Civilian Industrial Technology effort in three industries: textiles, shoes, and building materials.[14] The program was a conspicuous failure, in large measure because the industries it sought to help had not asked for help, and objected to what they saw as government interference.

Another type of government selection of a specific non-defense technology for technology support is the large-scale demonstration project. The "demonstration" is launched as a one-time project, ostensibly to test both technical and market risk. In most cases, however, little risk of either kind could be tolerated, given the huge public expenditures involved. Among the most visible examples are the Supersonic Transport, the Clinch River Breeder Reactor, and the Solar Photovoltaic, Oil Shale, and Coal Gasification projects during the energy crisis of the mid-1970s. Demonstration projects often arose as a compromise between those who urged direct government aid to a specific industry whose health was deemed critical to the national interest, and those who favored free market forces over government subsidies.

Linda Cohen and Roger Noll have showed that such projects are often captured by their constituencies and fail to adapt either to market or to technology change.[15] As the Council on Competitiveness put it, "Some fear the United States is too preoccupied with national prestige technology projects to worry about investing in the generic enabling technologies that are critical to the competitiveness of many industries."[16] Sylvia Ostrey notes that conservatives decry the tendency of governments to "pick winners and losers," and observes that "winners are very good at picking governments."[17] In the latter part of the Carter administration, it became evident that such projects were rarely of substantial help to the industries concerned, and they began to fall from favor. The fall became precipitous in the early years of the first Reagan administration, when virtually all civilian technology projects were cancelled, and funds were shifted to military development and basic academic research.[18]

Activities such as commercial demonstration projects and the Civilian Industrial Technology program laid the groundwork for interested parties to advocate the merits of specific industries and their core technologies as "critical" to the national interest. David Guston notes that many technologies are specific to a given industry, and may be understood politically as "crystallized interests."[19] But the political controversy that accompanies the identification of "critical technologies" with vested interests led government officials to look for alternative policies that were not overtly technology-specific. Thus, throughout the Cold War period, both technology specific and technology non-specific policies were initiated.

The next chapter discusses alternative policy criteria for federal investments in commercial technology, and discusses the agencies—existing and proposed—for carrying out this mission. One of these programs, slated for expansion by the Clinton administration, is the Advanced Technology Program (ATP), managed by the National Institute for Standards and Technology (NIST) in the U.S. Department of Commerce. The ATP provides cost-shared funding to commercial firms, or consortia of firms, in support of commercially relevant technology. As currently administered, however, the program does not give priority to any list of "critical technologies," even though NIST provided such a list two years ago. Instead, the program is open to unsolicited proposals from firms for work in what the Commerce Department calls "high-risk, high-return research on pre-competitive, generic technologies."[20] The selection process serves up projects that do not *a priori* follow any list of pre-determined critical technologies, thus avoiding the prior choice of technologies to be funded. In this way the ATP avoids appearing to select specific industry "winners."

In the aftermath of the Cold War, *de facto* U.S. technology policy has been shifting rapidly from its emphasis on development of technologies to serve missions assigned to the federal government (particularly in the defense, space, and nuclear fields), to programs intended to enhance the economic performance of private industry. (See Figure 2-1.) While most of the technology-specific interventions summarized in the table are related to a defined public interest (for example, transportation, or domestic sources of en-

Figure 2-1 Examples of Postwar Civilian Technology Policy Initiatives

Technology-Specific Initiatives	Generic or Indirect Initiatives
1946: Atomic Energy (nuclear power & medicine; Atomic Energy Commission)	1979: Carter Domestic Policy Review (generic technologies; all agencies)
1965: Air Transportation (super-sonic transport; DOT)	1980: Stevenson-Wydler Act [PL 96-480] (universities & national labs–industry cooperation; all agencies)
1968: Civilian Industrial Technology (CIT) (shoes, textiles, building materials; DOC)	1980: Bayh-Dole Act [PL 96-517] (privatize government patents)
1971: Project Breakthrough (factory-built housing; HUD)	1981: R&D Tax Credit
1975–78: Alternative energy sources (oil shale, coal, solar; DOE)	1984: Cooperative Research Act of 1984 (industry R&D consortia; all agencies)
1979: Civilian Automotive Research Program (CARP) (autos; DOE)	1986: Federal Technology Transfer Act of 1986 (national lab–industry collaboration; all agencies)
1990: Global Environmental Change (FCCSET review; multi-agency program)	1988: Omnibus Trade & Competi-tiveness Act of 1988 (Advanced Technology Program & Industrial Extension; NIST, DOC)
1991: Surface Transportation (Intelligent Vehicle Highway System [IVHS] in DOT)	1992: Small Business Innovation Research (SBIR) amendments (doubling SBIR budget; all agencies)
1991: High Performance Com-puting & Networking (HPC Act of 1991)	

ergy), technologies aimed at economic revival have no *a priori* criterion for choice.[21] A process must be devised to rank in priority order the specific civilian technologies to be supported if a technology-specific strategy is to be pursued for economic revitalization. This mechanism has, since the late 1980s, been the Critical Technology List.

Post–Cold War Critical Technologies: Activities and Legislation

In 1987, a private trade group, the Aerospace Industries Association, published a study made by many panels of experts from industry, universities, and government laboratories, identifying a number of technologies the industry considered "key" to its success.[22] The Association, headed by Donald Fuqua, retired chair of the House Science, Space and Technology Committee, wanted to convince the government that the industry it represented had both commercial and military value, and that in promoting economic performance the government should support technology development as well as basic research. Although the technologies it analyzed were called "key" rather than "critical," the study was a prototype for those that Congress would soon ask for.

The computer industry was not to be left behind. The Computer Systems Policy Project (CSPP) comprises the chief executive officer and the chief technical officer from each of the eleven largest U.S. computer firms. Its 1990 report *Perspectives: Success Factors in Critical Technologies* identified 16 critical technologies for the future, and discussed the success factors that might guide public technology policy.[23]

During the first Reagan administration, the Department of Defense began to express its concern about the state of the civil economy and its future impact on defense, should the decline continue. The Defense Science Board, in its study of U.S. dependence on foreign sources of supply for semiconductors (which later led to government support for the SemaTech consortium), worried about the impact of lagging U.S. competitiveness on defense interests.[24] The Undersecretary for Acquisition and Logistics, Robert B. Costello, proposed that the Defense Department invest in the civil industrial technology base on which defense production technology depended. These concerns energized Senator Jeff Bingaman (D-NM), a member of the Senate Armed Services Committee and the Joint Economic Committee. As a result of his efforts, the National Defense Authorization Act for FY 1989 required the Defense Department to provide Congress annually with a Critical Technologies Plan.[25] The statute defines critical technologies as "the technologies most essential to develop in order to ensure the long-term qualitative superiority of U.S. weapon systems." The criteria given in the report for selecting technologies for the list were: technologies that enhance performance of conventional weapons and provide new military capabilities, and technologies that improve availability, dependability and affordability of weapons systems.[26] The first such report, listing 22 technologies, was sent to Congress on March 15, 1989.[27]

The 1990 Defense Critical Technologies List added two more criteria: pervasiveness in major weapons systems, and strengthening the industrial base to "reflect explicitly the growing concern for spin-off to the industrial base."[28] However, as shown in the next section, these criteria could apply to the full range of defense R&D investments, and thus fail to provide much guidance or constraint on selection of candidate technologies for government investment.

The Technology Administration of the Department of Commerce offered its own analysis in 1990, calling its list "emerging" technologies.[29] Its basis for selection was the potential of the technology to contribute substantially to the economy over a ten-year period. The criteria included: potential market size, potential contributions to productivity or quality improvement, and potential to drive next-generation R&D and spin-off applications.

The FY 1990 Defense Authorization Act allocated defense funds to finance a Critical Technologies Panel to be established by the Office of Science and Technology Policy (OSTP). It would report biennially to Congress on technologies critical to meeting national needs: competitiveness, defense, energy security, and quality of life. Because any OSTP study of critical technologies might heighten pressure on the administration to move toward what it felt was an unacceptable form of "industrial policy," the Bush White House was quite nervous about the Panel's work. When the Panel's draft report appeared in the spring of 1991, the *Wall Street Journal* reported that the "White House Tries to Distance Itself from Panel Report"; the article implied that the report was indeed a step down the slippery slope of industrial policy.[30] The pressure became quite intense; OSTP issued a press release disavowing any such intent, and all identification of OSTP or the Executive Office of the President with the report was deleted. The President's Competitiveness Council, chaired by Vice President Dan Quayle, issued a Fact Sheet on April 25, 1991, acknowledging the Panel Report and noting the importance of technology creation, while stating that, with certain exceptions, administration policy was to leave this task to the private sector.

The second report in this series, prepared by a new panel under the guidance of OSTP with the help of the new Critical Technologies Institute, was completed in December 1992 and was published in 1993.[31]

The private sector Council on Competitiveness published a report in March, 1991, entitled *Gaining New Ground,* that assessed the competitiveness of U.S. industry in nine technologies, representing over $1 trillion in sales. "The sole criterion for choosing the...technologies is their importance for the competitiveness of the industrial sectors studied."[32] This listing of critical technologies has the virtue of being less arbitrary than many of the others. Panels of experts examined each of the nine largest industrial sectors without prior assumptions about the value of the sector to government or the national interest, other than the potential to create wealth or employment. The report recommended a five-year national program of investment in "critical generic" technologies.[33] (Its other recommendations deal with both diffusion strategy and R&D unlinked to the specified critical technologies.) Thus this report, unlike most of its antecedents, does specify what is to be done with the analysis, rather than leaving the issue open.

As the result of further initiatives taken by Senator Bingaman, the Defense Department budget for FY 1991 contained authorization for a Critical Technologies Institute (CTI).[34] The CTI was charged with responsibility for the biennial critical technologies panels established by OSTP, and funding provided in the first year for the CTI was used to this end. However, the Institute was not established until October of 1992 because the administration was reluctant to see it created and the Defense Department did not want to see its funds used for this purpose. A lengthy negotiation between Congress and Richard Darman, director of the Office of Management and Budget, resolved the impasse: the CTI would be a Federally Funded Research and Development Center (FFRDC), administratively established and serviced by the National Science Foundation.[35] A council chaired by the director of OSTP, with membership including three other senior White House staff, together with senior officials from S&T agencies listed in the statute, provides policy direction for the Institute. Funds for the Institute are transferred from the Defense Department to the National Science Foundation. The award was made to the RAND Corporation of Santa Monica, California, which operates the CTI at its Washington D.C. offices. To this extent, the critical technologies paradigm has been institutionalized in Washington.

Motivation, Goals, and Criteria for Making "Critical Technologies Lists"

The word "critical" carries little meaning to support policy; it clearly implies high value: that important strategic objectives may be unobtainable in its absence. Thus the loss of advantage in a "critical technology" might be thought a serious blow to the plans of its possessor. But criticality depends on context. Insulin is critical to a diabetic, but to no one else. The patent on the use of selenium as a photoconductor was critical to Xerox, until the day the patent ran out, but selenium would not be listed in the Defense Department's inventory of critical materials. A lithography system capable of patterning 0.35 micron circuit elements on a silicon wafer would be a critical technology to a microchip manufacturer, and probably to the nation whose industry enjoys strength as a microelectronics producer, but not to equally strong economies that are content to buy their chips from others.

Thus an economist, having calculated the economic return from a technology, might identify as critical the civilian technologies of greatest economic importance. A scientist will consider critical those tools without which his science cannot be explored. A historian might consider critical a technology whose introduction had the most profound effect in shaping the mores and institutions of society.

Corporations have diverse approaches to attaching strategic priorities to technologies. We can divide industries into two groups. Those most sensitive to monopolies based on legally protected intellectual property, such as pharmaceuticals and software, will consider as "critical" those patents or copyrights that preserve the opportunity to price according to product value rather than to cost. Those with broad ranges of technical choice will view "criticality" in terms of their business strategies. If the firm characterizes its business as defined by its core technologies, the method of selecting those technologies will define "criticality." The NEC Corporation, for example, has a planning methodology that selects from among its available technology futures the smallest number that support access to the broadest range of possible markets. A "market-defined" firm in the same industry, however, will select as critical to its future those technologies that, through diversification, best protect its markets from attack by firms from outside the industry.[36]

A military organization will consider critical those technologies that are not yet available to potential adversaries and provide the largest qualitative superiority to its forces, independent of cost. Because such advantages are transient, special importance will be attached to any set of tools, materials, or process equipment without which a weapons system providing qualitative advantage cannot be produced. Thus, economic return may be of minor interest. The highly specialized Toshiba machine tools sold to the Soviet Union, said to have enabled the USSR to substantially quiet its submarine fleet, would be an example.[37]

Former Deputy Assistant Secretary of Defense Jacques S. Gansler, author of *Affording Defense*, criticized the criteria used by the Defense Department in its critical technologies report to Congress and proposed the following criteria, in an attempt to find criteria more explicitly related to priority for defense investment:[38]

•Defense test: essential for future military equipment

•Technology manufacturing test: high barrier to entry in manufacturing

•Reconstitution or surge test: stockpiling is not a practical alternative

•Vulnerability test: multiple sources of supply could not provide an alternative

•Linkage test: technology is widely used or fundamental; one-time purchase is not a practical alternative

•Substitution test: no substitute is available

•Government leverage test: there are practical steps that government can take to preserve access to the technology

Gansler noted that such criteria not only shorten the "list" but also provide protection against political distortion of resources allocated to critical technologies development. As this example suggests, any set of criteria for "criticality" must be specific to the goals and purposes to which the technology is to be applied.

We argue that a government developing technology policies to sustain the competitiveness of its commercial industry should have its own specific criteria for defining a "critical technology." It should ask into:

•The economic value of a technology: What opportunities for job and wealth creation in domestically located industry are inherent in the technology? How widespread are the economic benefits from a leadership position in the technology?

•The opportunities for differential advantage: Is there enough technical challenge in the technology, or enough opportunity for intellectual property protection, to provide first-mover advantage to the successful innovator?

•The opportunity for a government to influence progress: Would actions by government, such as investing in R&D, targeting procurement, setting standards or providing financial incentives, influence the rate of commercial progress?

•The appropriateness of federal action: If the answers to the first three questions are affirmative, there remains the question whether such government action is appropriate, under economic theory for a competitive, market economy and under fairness principles in a democracy.

The key issues controlling the decision to intervene in the economy, spelled out in detail in Chapter 3, are:

•Does the technology, at its current stage of development, exhibit sufficiently low appropriability that private investment will be insufficient to advance it at a satisfactory rate?

•Does the technology have a high enough rate of social return to justify the public investment, either through economic return to the economy or through provision of public goods?

None of the eight U.S. studies that have led to lists of critical technologies have in any formal sense used any such set of criteria. Thus, these studies do not support the conclusion that government should necessarily invest in the technologies on the list, nor does the existence of such a list suggest what actions government would be justified in taking.

Problems with Using "Critical Technologies" to Allocate Resources

Among the many problems of relying on the adjective "critical" as a criterion for allocating resources is that it has become devoid of real meaning. Examination of the criteria for listing a technology as "critical" illustrates this most basic shortcoming of the label. Most of the technologies said to be critical are at best important opportunities for the private sector, but it cannot be said of every one of them that the failure of private industry to practice it successfully threatens the survival of the nation. The United States enjoys a lead in the formation of new firms using biotechnology to create new pharmaceuticals and other products, largely thanks to over $60 billion spent by the National Institutes of Health on biotechnological research in the last twenty years. It should, therefore, come as no surprise that the Critical Technologies Panel would include biotechnology on its list. But including biotechnology in a U.S. list is only documenting the obvious; biotechnology receives the lion's share of both the federal government's basic research dollars and of venture capital. Success is critical to government managers and venture capitalists; only in that sense is the technology "critical."

One problem with recent efforts to set priorities for resource allocation to specific technologies is that they tend to be defined at such a high level of abstraction that all lists look similar, and the identification of a technology area such as "robotics" or "biotechnology" is so broad that it provides little guidance to program managers. Mary Ellen Mogee extracted from these reports a list of the technologies each included in its analyses.[39] The matrix thus formed is remarkably dense; all the reports tend to pick the same technologies, for example:

•Manufacturing: Computer Integrated Manufacturing (CIM), robotics, artificial intelligence

•Information technologies: high-performance computing, displays, signal processing; human factors, networking

•Biotechnology: medical applications

•Aerospace: propulsion, surface transportation systems

•Materials: materials processing, composites, optoelectronics, microelectronics

The main differences among the critical technologies reports result from the fact that the private sector reports restricted themselves to their own industries. The defense list contains a number of highly specialized weapons technologies of very limited application, such as hyper-velocity projectiles and rail guns. The national Critical Technologies Panel list is broadest, including energy and environmental technologies.

All of these reports focus on the very areas that have already attracted the most commercial and government attention. The lists developed by foreign governments are also very similar. This tendency may reflect the desire of specific industries to see government contribute to their core technologies and of government agencies to protect their existing programs and budgets. But it leads government to invest in the very technologies to which the private sector has given priority. The resource allocation process should, instead, be flexible enough to identify new and exciting opportunities that might not be on someone's list and in which the private sector underinvests because of some market failure or a lack of vision.[40]

Need for a Distinction between Government and Industry Roles

Whatever attributes are associated with the term "critical," they tend to convey a sense of priority. A list of critical technologies specifies the priority attached to different technical objectives, but fails to make a distinction between the role of the federal government and that of the commercial or education sectors or the states. Nor does "critical" define the appropriateness of a government investment in private sector R&D. Appropriateness depends on three elements: (a) the fairness of the process for selecting a private organization to benefit from public funds; (b) investing in public goods, not merely the avoidance of substituting public for private funds; and (c) addressing market failures, including the failure of private firms to address early stage technologies of potential value.[41] Fairness is a matter of process, while restricting government support to public goods rests on economic criteria. These economic criteria relate to market failures, such as technologies with low appropriability to the firm but potentially high social return.[42]

In the search for simple concepts that capture appropriateness and fairness, Congress and the administration have used terms such as "pre-competitive" to suggest non-distortion of the market, "generic" to suggest ubiquitous utility, and "enabling" and "infrastructural" to imply indirect contributions to productivity and to the innovation process.

"Critical" Fails to Embrace the Sense of Generic Technologies

The National Advisory Committee for Aeronautics (NACA), parent agency to NASA, gets high marks, especially during the period between World Wars I and II, for providing what today would be called generic and infrastructural technologies in support of the growth of a young aeronautics industry. Much of what NACA did was enabling—building wind tunnels, devising test methods and safety standards, providing a common meeting ground for the industry's technical experts. None of this would be included in current lists of "critical technologies." Indeed, had the NACA work been on the "critical path" to any one firm's new aircraft development, there would surely have been complaints from competitors.

The research work of the National Institute of Standards and Technology (NIST) is highly regarded by industry as contributing to productivity improvement and reducing transaction costs between buyers and sellers of technology. But again, none of it can be considered critical. Notwithstanding this concern, the Council on Competitiveness chose to embrace the oxymoron "critical generic technologies."

Discounting the Importance of Technology Diffusion

Although the better reports (such as that of the Council on Competitiveness) do give emphasis to enhancing access, adaptation, and utilization of technology, legislation dealing with critical technologies is overwhelmingly biased toward the *generation* of technology as the role for the federal government. Yet post–Cold War technology policy in the United States should place much greater emphasis on technology *diffusion* to and within the private sector, as explained in Chapters 4 through 8.[43] Most of the technology that private firms need to improve their performance is already developed. Industrial extension services (Chapter 6), more fully developed information infrastructure (Chapter 5), and human resource development (Chapter 7) are the key investments required.

Need for Comparative Evaluation of Industrial Performance

Any government that seeks to bolster industrial competitiveness by investing in commercially relevant R&D must begin with good information about the competitive strengths and weaknesses of its industry and the nature of the opportunities and the areas of economic stress or risk. Such an analysis produces a net technological assessment, industry by industry. Such information is prerequisite to setting priorities in government industrial policy. In order to perform this analysis, government must choose between statistical economic analysis and comparative evaluation of specific technological capabilities. Most economists would insist that statistical analysis by industrial sector, without an arbitrary choice of the sectors to be studied, is the only justifiable way to assess overall performance of the domestic economy. However, this approach reflects past technical comparative advantages, whereas what is needed is predictive value to illuminate policies for future action. In order to make such an analysis, one must first identify the technologies in each industrial sector that are judged to be most critical in their potential contribution to competitive advantage. Technologies in which U.S. firms are falling behind can then be identified. Pending legislation (H.R. 820 and S. 4) would authorize the Department of Commerce to establish just such a capability.

Dangers of Coupling Promotion with Protection

As discussed above, the criteria for criticality for export control purposes begin to resemble those one might apply in a policy designed to enhance commercial competitiveness. Indeed, the first Congressionally mandated "lists" were required of the Department of Defense, and the criteria stated in the department's report to Congress read like those for an export control process. So long as the policy consequences of identifying a "critical technology" are either unspecified or are ambiguous, such lists might become instruments of trade protection as well as a guide for domestic technology promotion, causing inevitable conflicts.

Setting Priorities for Government-Supported Technologies

One reason the subject of critical technologies has attracted so much political attention in the United States is that four sectors of the U.S. science and technology enterprise are candidates for the mission of generating "critical technologies": private firms, government cost-sharing with firms, government-funded national laboratories, and universities with funding from both government and industry. If there is a "new" mission in critical technologies in the modern industrial state and public funding is involved, there will be competition for the role of generating critical technologies. If such new policies are to be implemented, they must be clearly stated, with objective criteria for project selection and evaluation that are explicitly related to the stated goals. Selection criteria must differentiate the government's role from that of private investors and entrepreneurs, and set priorities for government investments in civilian technology in ways that are resistant to interest-group capture.

How, then, are those priorities to be set, and the supporting analysis to be performed? If one takes a critical technologies list as the starting point for the analysis, economists will criticize the choices as arbitrary, since the relationship between important technologies and economic outcomes is unclear. An alternative and more rational starting point is to reverse the process and begin with an economic analysis of all economic sectors to find those with the best growth potential, then look at what sectors depend on performance in technology, and finally examine within those sectors the leverage that any particular technologies might have.

This approach is much less arbitrary, but it still has several shortcomings: It will not reveal nascent industries still in the venture capital mode, in which technology may be decisive (as biotechnology was in 1985). It will not identify new technologies that may in the future transform a stagnant industry into a rapidly growing one (e.g., optical fiber cable, for a firm that had made copper cable). It will not identify the infrastructural or generic technologies whose impact is spread across many industrial sectors.

A strategy based on both technical judgment and microeconomic analysis may be needed. There are two ways to develop the strategy. One approach starts by asking a broad cross-section of industrial technology leaders to identify highly leveraged technologies, and uses microeconomic analysis to assess the consequences of competitive standing in those technologies. Alternatively, one might start with the census of manufacturing, identify the industrial sectors with the greatest dependence on technology and potential for growth, and then engage experts in international industrial technology to assess the condition of the national industry and the potential for competitive advantage.[44] Both approaches should lead to similar results.

The idea that programs meriting priority will be called "critical" will have to give way sooner or later to the reality that the government's role is not defined by "criticality" but by a set of more specific criteria for government investment, such as those described in *Beyond Spinoff* and discussed in Chapter 3. Such criteria as "pathbreaking" and "infrastructural" are not technology-specific; within each category many technologies may compete for attention. What might be top priority in a list of pathbreaking technologies (high temperature superconductivity, for example) would probably not be considered critical as a generic technology. Indeed, the word "critical" seems ill-suited to a pathbreaking technology in which the government invests because the risks and delayed benefits do not make it an attractive object for private investment (and thus certainly not "critical" to a firm). Equally, the association of "critical" with "generic" fails to reflect the attribute of a generic technology that makes it appropriate for government investment: the generic technology benefits many if not all industries; it is "critical" to none. If, for example, reactive ion etching tools for making integrated circuits are seen as a critical strategic technology in SemaTech, that is because integrated circuit production has been deemed critical to the nation's well-being, not because of any specific attribute of reactive ion etchers.

Thus, a technology policy capable of defining the role of government in enhancing the nation's competitive advantage will suggest arrays of activities of different kinds. Within each type of activity, priorities will have to be set. Lists of critical technologies must then be expressed as multidimensional arrays.

Signs that the policy debate is becoming more sophisticated are at hand. Although Title III of the American Technology and Competitiveness Act of 1992 is entitled "Critical Technologies," it focuses not on lists of "critical technologies" but rather on "mechanisms to coordinate the development of a national policy, the implementation of which will ensure U.S. leadership in technologies (and their applications)...essential for industrial productivity, economic growth, and national security."[45] Priority setting for federal R&D, already launched by the Subcommittee on Research of the House Committee on Science, Space, and Technology, requires the development of clearer, more appropriate language in this area. The need for competitive and comparative assessment of U.S. industrial technology is clear. Maintaining a clear distinction between the use of export controls and export promotion should reduce the danger of confusion. The need to focus on the appropriateness of federal expenditures, and not just on priorities for choosing technical areas, should become more obvious.

Perhaps the best explanation for the emergence of critical technology studies is the recognition that government missions have defined technology priorities in the past, and that a new, more industry-centered way to set priorities in support of economic health must be evolved.

Notes

1. This chapter is based on a paper prepared for the Directorate for Science, Technology and Industry of the Organization for Economic Cooperation and Development. The views in the report are those of the author, not of the OECD or member states.

2. In 1988 West Germany devoted 14.5 percent of government-funded R&D to industrial development, Japan 4.8 percent, and the United States 0.2 percent. *Gaining New Ground: Technology Priorities for America's Future* (Washington, D.C.: Council on Competitiveness, March 1991), p. 14. The original source of the data is the OECD.

3. An amusing illustration of the distance between high science and economic performance is the strong inverse correlation between average real growth rate of GDP in 10 nations (1974–83) and the number of Nobel prizes won in physics and chemistry per capita. Christopher Hill and Joan D. Winston, *Science Policy Study Background Report No. 3*, Congressional Research Service study for the U.S. House of Representatives Task Force on Science Policy, Committee on Science and Technology, Ninety-ninth Congress, Second Session (Committee Print), Serial S, 192 pages, September, 1986.

4. David Teece, "Capturing Value from Technological Innovation: Integration, Strategic Partnering, and Licensing Decisions," in Bruce G. Guile and Harvey Brooks, *Technology and Global Industry: Companies and Nations in the World Economy* (Washington, D.C.: National Academy Press, 1987), pp. 65–96. Complementary assets are simply those assets or resources necessary for gaining economic return from a product or service, other than the ability to design, produce, distribute, and offer for it for sale. Examples might include software, training, low-cost components, and field service offered by others. In most cases of interest, the complementary asset lies outside the reach of the firm, although vertically integrated firms are often seeking the internalization of complementary assets.

5. CAT is Computer Aided Tomography. The device enables the making of three-dimensional pictures of the interior of the human body.

6. I am indebted to Harvey Brooks for emphasizing the importance of this point, which he has made convincingly in Harvey Brooks, lecture at The Herbert J. Hollomon Memorial Symposium, MIT, April 1991. The quotation is from Harvey Brooks, "The Relationship between Science and Technology," Nathan Rosenberg Festschrift Paper, October 14, 1992, to be published in 1993.

7. Richard R. Nelson, "National Innovation Systems: A Retrospective on a Study," *Industrial and Corporate Change*, Vol. 1, No. 2 (1992). See also chapter 16, Richard R. Nelson, ed., *National Innovation Systems: A Comparative Analysis* (New York: Oxford University Press, to be published 1993).

8. The traditional "linear" or "pipeline" model of the innovation process is initiated by basic research or invention, followed by exploratory development of a working protoype, called "upstream" technical activities. Then the prototype is given to the product engineering and production unit, whose activities, along with quality assurance and service, are called "downstream" activities. In practice the process is rarely as sequential as this model suggests, but may be initiated at any stage, from the market to the research laboratory, and is interactive.

9. Defense Science Board, *Report of the Defense Science Board Task Force on Export of U.S. Technology* (Washington, D.C.: Office of the Secretary of Defense) February 27, 1976.

10. Dual-use technologies are those applicable to both military and commercial uses. Post-Cold War U.S. military strategy is giving increased emphasis to command, control, communications, intelligence and logistics—all areas in which commercial technologies are strong and defense technology investments are dual-use.

11. Export Administration Act, P.L. 96-72.

12. The classified MCTL was first published in 1980 and was revised five times by 1986. The first unclassified version was published in 1984. See, for example, the second unclassified version: Office of the Undersecretary of Defense Acquisition, *The Militarily Critical Technologies List* (Washington, D.C.: Department of Defense, October 1986). The quotation is from p. ii.

13. Pre-competitive technology and other terms used in recent legislation to define the domain for public investment in commercially relevant technologies are defined and discussed in Chapter 3.

14. Dorothy Nelkin, *The Politics of Housing Innovation: The Fate of the Civilian Industrial Technology Program* (Ithaca, N.Y.: Cornell University Press, 1971).

15. Linda Cohen and Roger Noll, *The Technology Pork Barrel* (Washington, D.C.: The Brookings Institution, 1991).

16. Council on Competitiveness, *Gaining New Ground: Technology Priorities for America's Future* (Washington, D.C.: Council on Competitiveness, March 1991), p. 1.

17. Sylvia Ostrey, *Governments and Corporations in a Shrinking World* (New York: Council on Foreign Relations Press, 1990).

18. It should be noted that the Reagan administration was not consistently opposed to large technology demonstration projects, only to those aimed at commercial markets. The administration supported the National Aerospace Plane and many SDI prototype demonstrations.

19. Where technologies are specific to an industry, the making of a critical technologies list constitutes the designation of which firms, universities, and others with interests in public technology investments will be benefited. David Guston, private communication, December 30, 1992.

20. *Strategic View*, Technology Administration, U.S. Department of Commerce, November 1991, p. 25.

21. The two examples in the table that do not relate to infrastructure or energy, the CTI and the CARP, both failed on political grounds, suggesting that the economic interests of their advocates were not sufficiently crystallized.

22. Aerospace Industries Associates, *Key Technologies for the 1990s: An Overview* (Washington, D.C.: The Aerospace Industries Association, 1987).

23. Computer Systems Policy Project, *Perspectives: Success Factors in Critical Technologies* (Washington, D.C.: Computer Systems Policy Project, 1990).

24. *Defense Science Board Task Force on Defense Semiconductor Dependency* (Washington, D.C.: Department of Defense, Defense Science Board, February 1987).

25. Public Law 100-456, section 823.

26. Ibid., p. 5.

27. *The Department of Defense Critical Technologies Plan* (Washington, D.C.: The Department of Defense, March 15, 1989).

28. Mary Ellen Mogee, *Technology Policy and Critical Technologies: A Summary of Recent Reports*, Discussion Paper No. 3, The Manufacturing Forum (Washington, D.C.: The National Academy Press, December 1991).

29. "Emerging" is used in a sense similar to "pre-competitive." It represents a phase in the industrial life cycle rather than a priority *per se*, except that, as used in this context, it also carries the implication of great future potential for the industry. Thus emerging technologies are those that may lead to growth industries and therefore become "critical" to a society's strategic economic goals.

30. Bob Davis, "White House Tries to Distance Itself from Panel Report," *Wall Street Journal*, April 26, 1991.

31. *Second Biennial Report, National Critical Technologies Panel*, January 1993 (published for OSTP by the CTI, but because of political sensitivities no institutional sponsor or publisher is indicated on the document).

32. Council on Competitiveness, *Gaining New Ground: Technology Priorities for America's Future* (Washington, D.C.: Council on Competitiveness, March 1991). Note that three distinct bodies have similar, and thus confusing, names. The *Council on Competitiveness* is a non-governmental, not-for-profit group created by John Young, chairman of Hewlett Packard Corp. The *President's Competitiveness Council* was chaired by Vice President Quayle and may not survive the transition from Bush to Clinton. The *Competitiveness Policy Council* is chaired by economist C. Fred Bergsten. Its members were drawn from industry, labor, government, and the public interest, including a number of academics. Appointments are made within these categories by the president, the House, and the Senate. It was

created by Congress in the Omnibus Trade and Competitiveness Act of 1988.

33. "Critical generic," as used by the Department of Commerce, means technology essential to the industrial foundations of next-generation applied commercial technologies. In this usage, exclusive access is not essential. Some would say that "critical generic" is an oxymoron: a technology that is generic must be universally available; how can it still be exclusive and therefore critical? This attribute is what Robert White calls "its vertical nature." He sees critical technology as being upstream in the technological chain—a material, tool, or component used in downstream applications.

34. National Defense Authorization Act for Fiscal Year 1991 (42 U.S.C. 6686).

35. FFRDCs are defined in Chapter 4.

36. Lewis M. Branscomb and Fumio Kodama, *Japanese Innovation Strategies: Technology Support for Business Visions*, CSIA Occasional Paper No. 10 (Lanham, Md.: University Press of America, forthcoming 1993).

37. Jonathan Kapstein, "A Leak That Could Sink the U.S. Lead in Submarines," *Business Week*, May 18, 1987, pp. 65–66.

38. Jacques S. Gansler, "Restructuring the Defense Industrial Base," *Issues in Science and Technology*, Spring 1992, p. 56. See also Jacques S. Gansler, *Affording Defense* (Cambridge, Mass.: MIT Press, 1989).

39. Mary Ellen Mogee, *Technology Policy and Critical Technologies*, pp. 16 and 17.

40. See L.M. Branscomb, Testimony at hearing on H.R. 5231, the American Technology and Competitiveness Act of 1992, before the Subcommittee on Technology and Competitiveness, Committee on Science, Space, and Technology, U.S. House of Representatives, June 3, 1992.

41. I am grateful to Christopher Hill for pointing out that "in the thin markets typical of advanced technologies it can easily be the case that no one in the country is making the investments necessary to create the 'supply curve' for the technology." As Hill notes, this is a policy situation that frequently confronts small market-economy nations.

42. An excellent and well-documented review of the criteria for public investment in private technology is given by Harvey Brooks, "Towards an Efficient Public Technology Policy: Criteria and Evidence," in Herbert Giersch, ed., *Emerging Technologies: Consequences for Economic Growth, Structural Change, and Employment*, Symposium 1981 (Tubingen, Germany: J.C.B. Mohr [Paul Siebeck] Publishers, 1981), pp. 329–365.

43. Lewis M. Branscomb, "Toward a U.S. Technology Policy," *Issues in Science and Technology*, Vol. 7, No. 4 (Summer 1991), pp. 50–55.

44. MITI in Japan comes closer to having made such analyses than other industrialized democracies, but the MITI approach is much more strongly aligned to selecting "critical industries" than "critical technologies."

45. H.R. 5231 was sent forward to the House of Representatives by the Committee on Science, Space, and Technology in the summer of 1992.

3

Funding Civilian and Dual-Use Industrial Technology

Lewis M. Branscomb and George Parker

The magnitude of the problem is such that we cannot rely upon normal forces to maintain our advantage in technology. We are at the forefront in many technological areas. The costs of breaking new ground in some of these areas are high—higher than private companies or perhaps even private consortia are able to justify because the risks are so great....Other trading nations have recognized it in the area of civilian R&D and have taken steps to assist technological development. If we are to maintain our advantages in this area we must first of all accept the idea that it has become a proper sphere of governmental action.

Maurice Stans, President Nixon's Secretary of Commerce, testimony to the Congress, 1971.[1]

We cannot rely on the serendipitous application of defense technology to the private sector. We must aim directly at these new challenges and focus our efforts on the new opportunities before us, recognizing that government can play a key role helping private firms develop and profit from innovation.

William Clinton and Albert Gore, Jr., February 22, 1993.[2]

Because the government provides half of the nation's R&D investment and directs the purposes to which this R&D should be applied, the government exerts a significant—and direct—influence on private sector productivity and innovation. But the government's own missions—notably in defense and space—have in recent years consumed two-thirds of this federal R&D investment. As discussed in *Beyond Spinoff*, it is far from clear how much these activities contribute to U.S. competitiveness. Thus there are calls for radical change in the way the government manages its half of the nation's R&D capacity and the policies it pursues to enhance industrial competitiveness.

For twenty years the United States was so inhibited by an ideological aversion to discussions of industrial policy that it was unable to face up to the evidence showing a rapid decline in the nation's high-technology trade balance that began in about 1970. The emerging challenge to the global dominance of United States high technology industry was first recognized in the Department of Commerce as long ago as the late 1960s. Michael Boretsky, an iconoclastic, technology-oriented economist in the department, was tracking shares of world markets, segregating the U.S. economy into high- and low-tech segments. When he began, the favorable U.S. trade balance of high-tech exports was able to cover the accumulating deficits in low-tech sectors. Then he showed that the high-tech sector's world market share was slowly but steadily eroding, until in 1986 the U.S. high-tech merchandise trade balance went negative. Until then most government officials had felt that an adequate level of stimulation to industrial technology came from spinoffs from defense technology and university science, both of which were thought, incorrectly, to be both automatic and cost-free to the government.

Now the taboo on industrial policy has been lifted. The worsening trade and budget deficits have forced this issue to the top of the agenda and have emphasized the importance of leveraged public investments that, at minimum cost to government, enable firms to innovate more competitively. When it prepared the FY93 budget, the Bush administration had already started implementing many of the elements now found in the Clinton-Gore technology policy, even though Bush was unwilling to defend the policy very vigorously, much less claim credit for it during the campaign of 1992.

Despite what many claimants to a bigger share of the federal budget for R&D may think, the problem is not the adequacy of the total size of the federal R&D budget. In 1987, the U.S. government spent 60 percent more on R&D than did the governments of Japan, West Germany, France, England, Italy and Canada combined.[3] If the United States has a technological problem, it is because Americans are doing the wrong things and failing to make use of what we do. The issue is, "How do we harness the technical talent of America more effectively to the problems the country faces?"

A strong case can be made that the primary emphasis should be on demand-side policy: developing the human resources (see

Chapters 7 and 8), building information and technology infrastructure (see Chapter 5), and helping firms find, adapt and profitably exploit technology (see Chapter 6), most of which is available in the economy and was not created by the most recent federal R&D investments. This chapter examines how government R&D investments can create technology of value to the civilian economy.

Chapter 2 addressed the question of how technology should be selected for government support. In this chapter we address the next three questions:

•Under what circumstances is it appropriate for government to invest in commercially relevant technology?

•When it is appropriate for government to invest, how are activities in the selected technologies to be managed?

•What institutional innovations in government may be needed to implement the policy?

President Clinton has called for shifting $8 billion of defense procurement to civilian R&D investment.[4] What is the case for expanding the national investment in civilian R&D? In 1988, the U.S. public and private sectors spent 2.7 percent of U.S. Gross Domestic Product (GDP) on R&D, about the same percentage as Japan (2.6 percent) or West Germany (2.8 percent). This may suggest that U.S. R&D investments are adequate, at least by comparison with its competitors. But many people believe that the utility of most defense R&D to commercial industry is very limited. If that is so, it is civilian R&D, as a fraction of GDP, that we should examine. By this reckoning, the United States spent only 1.9 percent of GDP on civilian R&D, less than Germany and Japan by 0.7 percent of GDP. This is a "shortfall" of over $40 billion annually.[5] The United States should be looking to industry for the largest part of any increase in R&D investment. But there are many areas of R&D that are important to economic performance in which industry underinvests because of market failures or other disincentives. In these areas, government has a legitimate role in supplementing private investment with public expenditures on R&D.

How should such federal R&D funds be invested? If technology generation for the commercial industrial base is to be given priority, what is the strategy for justifying such investments, and for managing them, and what criteria are to be used to define and limit the role of government?

Policy Constraints on Publicly Funded R&D in Industry Partnerships

If Congress authorizes the expenditure of public funds on commercial activities, there must be some policy constraints to limit the perception that public funds are simply being diverted for private gain. These constraints, if built into a robust project selection process, should reduce complaints of unfair, government-distorted competition. They can also reduce the practice of political "earmarking," whereby members of Congress specify in appropriations legislation that funds for specific projects must be spent in their own districts or to benefit their own constituents. Political problems may also arise when federal funds are used to support commercial products from which particular companies or individuals will reap large financial rewards, unless it can be established that large spillover benefits to the public have also resulted. This issue is at the heart of the political debate over "industrial policy." Unless the public, the business community, and the government are all satisfied with how the responsibilities of the public and private sectors are distinguished, federal investments in commercially relevant civilian technologies will never be free of controversy.

There are two ways to approach the federal role in commercially relevant R&D. First, assuming that Congress may legitimately authorize funding of R&D for a broad range of public purposes, one asks what limitations must be imposed on such activities in the interest of effectiveness, fairness, and a competitive economy. This we call the "mission-justified" approach. Its starting point is that the government is a legitimate part of the market; program limitations derive not so much from theoretical limitations as from practical ones. Can the government do a good job, without undue political distortion, of serving the collective economic interest of the American people?

The other view begins with the assumption that government should never intrude into commercial markets except where market failures justify public investment. This *laissez-faire* argument has the advantage of ideological correctness, and the disadvantage that the burden of proof is on government for each and every intervention into the economy. The U.S. government uses both of these arguments, sometimes intertwined.

Mission-Justified Support for Economic Enhancement

During the past forty years the public has come to accept the legitimacy of huge federal investments in R&D in pursuit of the government's constitutionally granted responsibilities. No one questioned whether the Defense Department should pay for R&D to design weapons it intended to purchase. Once President Kennedy committed the nation to send American astronauts to the moon, it was clearly necessary to spend billions of dollars on R&D to allow the government to carry out its plan. In domestic policy, Congress has authorized agencies to develop peaceful uses of nuclear energy, to invest in battery technology for a future electric car industry, to pursue technologies for controlling pollution and disposing of waste, and a great range of other R&D activities.

In these cases the assumption is usually made that the desired activities will ultimately be carried out by the private sector. Such "mission-justified" programs have been the dominant form of federal stimulation of R&D of value to the economy in the Cold War period. The justification for such R&D is compensation for externalities the market does not adequately address. The constraint on the appropriateness of federal R&D investments, once Congress has authorized the program, is supplied by standards of effectiveness and fairness. Effectiveness is largely determined by the existence of a market (which government investments are often intended to help create) and the success at transfer of the technology from the R&D performer to the industry.

When the R&D is performed in government facilities, such as the national laboratories, barriers to technology transfer may be a serious obstacle to the effective commercial exploitation of the technology. (See Chapter 4.) This technology transfer obstacle is substantially reduced when commercial industry is the performer of the government-funded R&D. Fairness is then addressed by the use of competitive procurement, provided that there is a sufficiently active market to make competition effective.

The main limitation on the mission-justified approach is its restriction to a limited range of technologies accepted as legitimate for government investment. This is why listing "critical technologies" has become a focal point for policy discussion. The search for acceptable criteria for "critical technologies" is the effort to translate the legitimacy of mission-justified, categorical research and development to the world of the commercial technology base. (See Chapter 2.)

Government missions are tied to creating public goods such as safe roads, a clean environment, or cheap energy. A corresponding advantage of this mission-oriented approach is that there are natural constituencies for each of these missions—environmentalists, supporters of renewable energy, or auto workers and motorists, for example. It is harder to find the constituency for the goal of a more competitive industrial economy with reduced trade and budget deficits. The technologies that provide the civilian industrial base are not so easily defined; they cover the full range of modern applied science and engineering. The government may, therefore, be tempted to use the existing federal missions, which tend to be more narrowly technologically defined, to justify public investments to improve the competitiveness of an entire industry.

The Clean Car Initiative
The best example of such an effort is the Clinton administration's strategy for its Clean Car Initiative.[6] Advanced under the heading "Building America's Economic Strength," this proposal sets out to "facilitate private sector development of a new generation of automobiles." Thus the goal is unambiguously economic: to make the "auto industry...competitive and strong in the 21st century, preserving jobs, sustaining economic growth and expanding its business." But the source of legitimacy for this project is inherent in the technical goal: to help the industry develop "a new generation of vehicles...that would be safe and perform as well if not better than existing automobiles, cost no more to drive than today's automobiles, consume only domestic fuels such as natural gas and renewables, and produce little or no pollution."

While the goal of the proposal is to reverse the industry's loss of world market shares, the technical goal does not focus only on improving either of the qualities that matter most to car buyers:

performance and cost. Instead, the goal addresses economic externalities of the auto industry: lowering pollution, increasing public safety, and reducing the nation's dependence on imported petroleum. Success in addressing these three externalities may contribute to the collective well-being of the nation, but will probably increase automobile costs to consumers and tend to make the industry less rather than more competitive—unless, through luck, discoveries make it possible to reduce cost and increase performance while still reaching the societal goals. Thus the economic objective may be seriously compromised by the policy means of legitimation of federal expenditures. Nor is it clear that this plan contains the means to give U.S. industry a differential advantage over its foreign competitors. To the extent that the externalities are the primary motive for the program, the right policy would be to encourage world-wide industry to use the new technologies; to the extent that the goal is the competitiveness of U.S. manufacturers, tight control should be retained on the intellectual property.

Finally, one should note one more inhibition, in the interest of social objectives, on the likely economic effectiveness of the clean car project: the administration plans "a systematic search for capabilities in national laboratories and defense facilities" that can be of use in the project. This may help ease the burden of defense conversion on weapons laboratories and defense firms, but raises serious questions about the ability of those institutions to understand the automobile industry well enough to be an effective partner.

Small Business Innovation Research (SBIR)

A second example of mission-justified support for economic enhancement is the Small Business Innovation Research program (SBIR). The Small Business Innovation Development Act of 1982 requires eleven federal agencies to invest no less than 1.25 percent of their contract R&D in small businesses.[7] In 1992, $450 million in federal funds are estimated to have been spent in this program.[8] The Act does not authorize any new missions for the agencies; it simply requires that the agencies, in the course of their established missions, conduct a fraction of their work through small businesses.[9] Since many agencies may not find enough small businesses able to contribute in critical ways to their missions, they may be quite flexible in responding to small company initiatives so long as the proposal is a tolerable fit to the mission. In this way, mission-justified R&D is able to respond to the commercial interests of small businesses across a very broad range of industries. The program is specifically designed to encourage commercialization of the R&D they do for the government.

During the 1992 presidential election campaign, both Bill Clinton and George Bush advocated doubling the SBIR program. On October 28, 1992, a week before the election, President Bush signed Public Law 102-564 (the Small Business Research and Development Enhancement Act), which continues the SBIR program to the year 2000, and in stages will double its size to 2.5 percent of each agency's contract R&D. A "small business" is one with fewer than 500 U.S. employees, with at least half of the firm's stock held by U.S. citizens. In fact, firms with ten or fewer employees have won more than one-third of all awards. The maximum allowed award will, over a period of a few years, be doubled to $100,000 for initial grants and to $750,000 for follow-on grants that support commercialization. Patents are assigned to the firm for not less than four years after Phase II completion. The legislation also includes a number of measurements and assessments, primarily aimed at testing the program's success at commercialization, that the program must survive if the award size increases are not to be curtailed.

Exceptions to *Laissez-Faire*: Legitimate Subsidies for Commercial R&D

Given the deeply held conviction in the United States that the government should not attempt to do industry's job, it would be strange if the government's priorities for civilian R&D matched those of private firms. The government should, therefore, approach the funding of commercial technology with some caution, and should look for complementary roles, investing in technologies industry may need but cannot afford because single firms cannot capture enough of the benefits.

Historically, American economic rhetoric has adhered strictly to *laissez-faire*, even as Congress from time to time created efforts to remedy perceived problems in one economic sector or another.

These quite conscious exceptions to stated economic policy have been sufficiently unsuccessful to reinforce the *laissez-faire* rhetoric.[10]

In 1979, President Carter sent to Congress proposals to "significantly enhance our nation's industrial innovative capacity and thereby help to revitalize America's industrial base."[11] His proposal included funding for "generic" R&D of value to industry. This initiative stalled under Reagan, but was implemented by Congress and signed by President Bush in the 1988 Omnibus Trade and Competitiveness Act. The basis for this kind of investment (which has been characteristic of much of the research in laboratories of the National Institute of Standards and Technology (NIST), the Geological Survey, and other agencies for more than 50 years) lies in economic theory.

Economists recognize that where there are demonstrable market failures it may be appropriate for government to compensate for underinvestment by private firms. A common kind of failure is a lack of appropriability of returns to a firm investing in technology, which, when diffused throughout industry, nevertheless provides social returns substantially in excess of cost. Examples of the most frequently encountered market failures, discussed in detail below, that may justify government investment are:

•Basic research in areas of science supporting a firm's core competencies and exchanged freely with scientists and engineers in industry and universities (e.g., chaos theory, mathematics)

•Pathbreaking technologies with the potential to create new industries or product clusters, but with high risks and long payback periods (such as fusion power, electric cars, high temperature superconductivity)[12]

•Infrastructural technologies and technologically motivated science with broad application throughout industry in which the payback is relatively certain and rapid, but whose economic value is only weakly appropriable to any single firm (such as design-automation tools, materials characterization, properties of matter and materials, non-destructive test methods, and physical and chemical data compilations)

•Technologies serving markets with high barriers to entry, such as those whose entry costs exceed what private capital markets will support, given the extended time required to recover the initial investment (SST, MagLev trains, etc.)

•Strategic technologies, where markets are too small or too cyclical to sustain the technological infrastructure required for viability of an important domestic industry (such as the semiconductor manufacturing tool makers supported by SemaTech)

•Technologies for mixed public/private markets, where public policy issues have the effect of increasing private risks (such as molecular biology and the intelligent highways system), or where public values justify an acceleration of development effort beyond what the market will elicit (such as environmentally superior processes)

Basic Research

Since the end of World War II, conservatives and liberals have accepted the legitimacy of federal government investment in basic research. The policy basis for support of university research has rested on a tacit social contract, which delegates to the scientific community the initiative to define promising research objectives and to invoke merit review of competing proposals as the mechanism for allocating basic research resources. In return, the scientific community assures the society that benefits far in excess of cost will flow from the aggregate investment, even though the economic value of individual projects cannot be assessed. This social contract is under heavy pressure today, as Congress asks the scientific community to demonstrate the effectiveness of the mechanisms through which those benefits flow. (See Chapter 7.)

Basic research in the universities certainly enjoys a deeply committed constituency, but its political strength is sapped by the very social contract that establishes the federal commitment. Scientists are aware that they are a privileged group, not only because they constitute a highly educated elite, but because they enjoy such a high degree of autonomy. Few scientists are willing to lobby for what is obviously a self-serving cause.

Pathbreaking Technologies

Daring investments in pathbreaking technologies with the potential to create new industries can give Americans first-mover advantages. A balance of input from scientific vision and from assessments of industrial potential should guide choices of such investments. Public investments in basic science have traditionally been justified by the propensity of new science to give rise to new technology. In the current world of fast-paced commercialization, however, the first discoverers of a new phenomenon with industrial potential must move quickly to explore the materials and processing technologies required to exploit it, if the discoverer expects to enjoy the economic rewards of the discovery. When superconductivity at high temperatures was discovered by two IBM scientists (Nobel laureates Muller and Bednors), the Japanese moved very quickly to reproduce the science. At the same time, MITI built a consortium of companies to explore materials processing and other technical issues that would give them a lead if the discovery proved to have commercial potential. In this particular case the applications potential has so far proved elusive. But it is a feature of pathbreaking technologies that a great deal of patience is required. Thus the prospect of attractive markets years in the future is insufficient incentive for private investment in pathbreaking technologies, especially in fields where patent protection is likely to be weak. Nevertheless, once the vision of that future has crystallized, a strong constituency of committed believers, both scientists and business visionaries, will keep the vision alive.

Historically, the most committed constituency for pathbreaking technology has been the Department of Defense. The Defense Advanced Research Projects Agency (DARPA, now ARPA) has proved itself adept at identifying and supporting pathbreaking technologies. Its criteria for selection (and the risks to private investment) are primarily technical, once the military potential is identified. The strategy for funding this kind of work must rest on completing the theoretical understanding of the phenomenology and on the identification of the most critical barriers to commercialization. Government funding of the technology effort (but not of the basic science) should be terminated when the new industry takes off and no longer under-invests, or when it becomes clear that the vision of a new industry is a mirage. The process for making those evaluations must be built into the process for initiating funding.

Infrastructural or Generic Technology

Investments in infrastructural or generic technical knowledge may be relevant to the downstream concerns of industry and the need to increase R&D productivity. The Department of Commerce's National Bureau of Standards (now the National Institute for Standards and Technology, NIST) has been generating this kind of useful technical knowledge for industry for over 90 years, but it has conducted the work in its own laboratory on too small a scale to have a major impact on industrial performance. Examples of this kind of research are the characterization of materials and materials processing, research on design automation, simulation and modelling, development of new tools and instruments, and the determination of the properties of matter and materials.

The form in which the knowledge is produced and communicated is important to its technical and economic value. Consider a hypothetical example: a firm making micro-electronic packaging suffers a process failure in a production line because of yield-destroying cracks in indium-copper metallurgy. The engineers need to know the rate of diffusion of indium in copper in the presence of electric current at elevated temperatures. An alternative alloy metallurgy does not seem to exhibit the failure, but its performance is degraded. Researchers look in the literature and can find no data on diffusion rates of indium in copper under these conditions. The project manager has a choice: take a six-week delay and ask the corporate research lab to measure the diffusion rate, or switch to the inferior alternative metallurgy, and accept lower quality or poorer performance in the product.

A firm will almost always take the second alternative. Even if it does not, it will make measurements only at the concentration ratios, the current densities, and the temperature range relevant to solving its immediate crisis. The data will probably never be published; if it is published it will be a fragment of information that is hard to find and evaluate. This is a classic example of applied problem-solving research. The fragment of knowledge produced to meet the need is so narrow, so unlikely to be well publicized and so useless as a source of theoretical understanding as to have high appropriability and very low externality in its social return. Government should not fund this kind of work.

What society does need, on the other hand, is support for a scientist who is fascinated by the theory of electromigration in alloys, who measures several dozen alloys over a broad range of currents and temperatures, develops a theory, validates its predictability and publishes both his data and a computer model from which hundreds of other like situations can be estimated. Ideally, she works in a university; her doctoral students may take jobs in industry and take with them the embedded knowledge of how such alloys behave. Her intuitive judgment, together with the theory and data, can be a powerful asset in the form of good technology. In the first example, where the measurements are made in the problem-solving context, the appropriability of the work is high and immediate, but the rate of social return is low. In this example, the economic value is higher, and the benefits are distributed over a large number of potential users of the knowledge and over a considerable period of time. Accordingly, this systematic research to build a comprehensive base of materials knowledge is more worthy of public support, and is appropriately considered a contribution to knowledge infrastructure.

At the heart of the idea of infrastructural or generic science and technology is the aggregation of an organized knowledge base, structured and critically reviewed to insure both accessibility and rigorous quality. The international basic research community has long understood that data collection and evaluation, well-verified mathematical models of natural processes, and tools and instruments are the essential requirements for the efficient advance of scientific knowledge. For the efficient advance of a high-tech economy, the nation must support the development of well-characterized materials and processes, automated tools for design, simulation and testing, engineering handbooks, quality control procedures, production process tools and controls, and systems engineering knowledge for planning and executing complex programs.

Infrastructural or generic technology faces serious constituency problems. While the work is often quite basic, even theoretical, it rarely touches the conceptual frontiers of science. Thus it is too practical and too useful to earn high prestige among academics. At the same time this work is too systematic and too oriented to intrinsic scientific goals to thrill an industry R&D manager. The subject matter covers the full spectrum of industrial know-how, so it is even difficult to find a disciplinary constituency to support it. Infrastructural technology research is, however, a vital service to the world science and technology enterprise. To sustain it at adequate levels requires the development of institutional structures dedicated to the purpose. The NIST laboratory, where the technical culture celebrates infrastructural research, presents that intellectual environment. But comparable work should be supported by government grants and contracts so many institutions can contribute.

Technology with High Barriers to Entry
Where the entry costs exceed the risk tolerance of private firms and constitute an unreasonably high barrier to market entry, federal commitments to fund early-stage technology may be required on occasion. The familiar example is the Supersonic Transport project. Very large-scale commercialization projects of this kind have been discussed in detail by Cohen and Noll.[13] They present special problems because their constituencies become excessively strong and the projects are susceptible to capture. The barriers, however, may also be regulatory, as in biomedical research.

Strategic Technology
Occasionally government may be justified in making investments in selected strategic industries suffering from foreign targeting, with the goal of bringing manufacturing back to the United States so that the jobs are created here. The problem is not a market failure so much as a structural problem that may be the result of lack of foresight by industrial managers, or the failure of government to ensure fair international trade practices, or both. But when the decline affects a significant national interest (such as defense) and is seen as potentially irreversible, government action may be warranted. SemaTech (the Semiconductor Manufacturing Technology Consortium) seems to be a successful example. Funded in equal parts by government (DARPA) and by the U.S. microelectronics manufacturing industry, SemaTech sets common standards for interfaces to the production tooling, and provides R&D support to small and seriously threatened manufacturers of that tooling.

78

The Clean Car Initiative also appears to be positioned as a strategic technology initiative, and it illustrates a critical shortcoming of the use of technology policy to reverse the decline of a major industry. While better technology would surely help, the auto industry's competitiveness problems are much more deeply rooted in the cost of labor (especially the cost of retirement and health benefits to workers), and in consumer perceptions of quality and performance of the cars. Thus for a strategic technology investment in the auto industry to achieve its economic goals, a broad range of non-technical actions may be required. A package of solutions to which the industry is committed should accompany any strategic technology investment.

Technologies for Mixed Public and Private Markets
Some potentially very attractive commercial ventures require coordinated public sector investments for their success. A current example is the Department of Transportation's Intelligent Vehicle Highway System, the goal of which is to increase the carrying capacity of American highways and roads by instrumenting the roads, allowing commercial digital telecommunications service companies to link a city or state's computers that are measuring traffic flows to communicate this knowledge to a computer in each car that guides its driver to the least congested route. The government assumes that automobile and communications firms will invest their own funds in these applications, but the firms must depend on city and state investments. In a situation of this kind, early cost-shared R&D is needed to prove out the combined applications so that both government and private firms can assess the risk and benefits of the venture and concert their strategies.

Program Designs

Crucial to the success of a civilian technology development program is an institutional design that reflects and reinforces the philosophy by which it is justified. The Clinton administration has declared a new willingness to commit federal resources to projects of commercial relevance. These programs must be able to foster the organizational linkages and interpersonal connections that are necessary for technology to be transferred from one organization to another.[14] A well-designed program acknowledges the complexity of the process of innovation: it involves not only feedback loops from manufacturing to research departments, but also complex vertical networks of suppliers and customers, horizontal networks among competitors, and networks among government scientists and corporate engineers, all linked by more and more sophisticated information technologies.[15]

George Heaton, in a far-sighted article written in 1989, suggested that a new paradigm for governmental involvement in commercial technology was emerging.[16] This new paradigm is characterized by two aspects: "First, the federal government is now committed to facilitate the development and deployment of commercial technology. Second, cooperative ventures among firms and/or sectors are an essential condition for public support."[17] Four years later, Heaton's vision is clearly becoming reality. It should be added that what is politically necessary is also practically beneficial. Several perspectives on innovation suggest that structures that encourage interaction among technical personnel greatly facilitate diffusion of new ideas.[18] Thus, cooperation plays a key role in the success of the projects, that is, in creating commercial innovation, not just in garnering public support.

Heaton points to three arenas of cooperation in the new paradigm: Public-private, interfirm, and university-industry. (University-industry cooperation is covered in Chapter 7.)

Public-Private Cooperation

In a well-designed program there should be an industry role in choosing, executing and funding projects. Since it is industry that has the ultimate responsibility to bring a technical product to fruition, any program that is to succeed in helping industry must be oriented toward industry's needs. There is no more effective way to do this than to have industry's input into the decisions that determine the choice of projects. Several of the existing programs that the administration wants to expand are already structured in this way. The ATP and SBIR programs respond directly to private sector proposals. The national lab CRADAs may also result from industry initiatives. Consortia such as SemaTech fund research reflecting their members' direct interests.

The requirement for a role for industry in developing a research agenda blunts the sharpest rhetorical sword of free-market critics. Critics allege that an active technology policy consists of the government "picking winners," and claim that government policies put the decision-making power in the hands of people removed from the detailed knowledge of the market. Clearly both of these objections can be met when the programs are developed by those who know the most about the market.

Industry involvement in agenda-setting raises two other objections, however: the programs may become mere pork barrel for industry; and the existing industry may fail to discern technological innovation that comes in quantum jumps.[19] The former point is addressed (but not resolved) by the requirement that the industry match the funds the government puts up. The latter point is met by noting that industry participation does not mean that *all* programs are developed this way: the realm of government technology development still includes support for basic research and for pathbreaking technology that is not yet targeted by industry.

Just as the decision of what technology to investigate should include private input, conducting the research also calls for private participation. Indeed the biggest problem of relying on spinoff or of depending on licensing government-developed technology is that the technology to be transferred is very rarely embodied in a device that can be simply carried from lab to industry. The intangible "know-how" (tacit or embedded knowledge) embodied not in a blueprint, but in a technician's head, is crucial to successful transfer of technology.[20] That is why the national laboratories' turn toward Cooperative Research and Development Agreements (CRADAs) is much more likely to result in commercial innovations than is the simple goal of licensing pre-existing technology, as conceived in the original Stevenson-Wydler Act of 1980. Transferring the inventor and the device is much easier than just transferring the device (which, in any case, does not exist in commercializable form until after the transfer is accomplished).

Cooperative activity should extend beyond project selection to funding as well. Shared funding is becoming the accepted mode for these new technology development programs. As President Clinton and Vice President Gore said, "the fundamental mechanism for carrying out this new approach is the cost-shared R&D

partnership between government and industry. All federal R&D agencies will be encouraged to act as partners with industry wherever possible."[21]

The Advanced Technology Program, CRADAs with the DOE national labs and with NIH, and SemaTech all require at least 50 percent private contribution. Shared funding leverages the public investment, shows a seriousness of purpose from the private sector, and helps sell the programs politically.

Interfirm Cooperation

The second arena of cooperation is at the interfirm level. A successful program will encourage or require firms in the same industries to form research consortia. By forming groups of similar corporations, the government will be acting as a catalyst in creating industrial clusters that are beneficial to long-term industry health.[22] These clusters have formed around the efforts of state-level technical extension agencies (see Chapter 6), and also as a result of projects such as MCC and SemaTech.[23] The cooperative nature of the groups also serves to force the research agenda away from the politically sensitive market end of research toward generic research that is less sensitive to competitive conflict. Having to cooperate to fund R&D may even lead competitors to fund their common suppliers, as in the case of SemaTech. An inclusive group of industry competitors also reduces the political problems arising from the have-nots. An additional way of meeting many of these political claims—the "winners" versus "whiners" problem—is to make opportunities open to all, within a system that is pluralistic, so that there are numerous venues for support within the system.

Firms which have participated in such cooperative groups often report extensive benefits from interpersonal contacts made and the informal (and formal) knowledge exchange that can help keep them on the cutting edge.[24] The benefits of cooperation may seem obvious, but it must be recalled that historically, in the American political economy, horizontal industrial interactions are not consistent with the individualistic, competitive American business culture, and are often fraught with anti-trust legal exposure. So, in practical terms, an official sanction may often be needed to get them started.

The National Cooperative Research Act of 1984 was the first legislative step in this direction, giving cooperative R&D legal sanction.[25] Legislation introduced in March 1993 would extend this protection to manufacturing consortia as well. The Clinton administration policy statement of February 22, 1993, proposes creating Regional Technology Alliances to "encourage firms and research institutions within a particular region to exchange information, share and develop technology, and develop new products and markets."[26] Also mentioned is an Agile Manufacturing Program to "allow temporary networks of complementary firms to come together to exploit fast changing market opportunities." While the historic resistance to cooperation may be difficult to overcome, these are the kinds of new programs that present opportunities for effective, new, loosely-coupled business institutions.

Management Issues

To be effective, a new technology development program should have clear criteria for selection of projects. Not all types of programs must have the same criteria; in fact they probably should not: infrastructure programs are different from strategic research consortia, which are different from university/industry projects. But all of the programs will have the most public support if they are judged to be effective and fair.

An interesting example is the Advanced Technology Program (ATP) run by NIST. Federal R&D cost sharing with commercial firms is justified by economic theory, but the way economic theory has actually been applied has created confusion. The Commerce Department's technology strategy states that the ATP program invests in development of "high-risk, high-return research on precompetitive, generic technologies." But "high-risk, high-return" technologies (called "pathbreaking" in *Beyond Spinoff* and in this book) are totally different from "generic" (or "infrastructural") technologies. Their purposes are different, the criteria for evaluation should be different, and the agencies responsible for them need different kinds of capabilities and different relationships with industry.

Generic technology is defined by the Department of Commerce as "a concept, component, or process, or the further investigation of scientific phenomena, that has the potential to be applied to a broad range of products or processes."[27] Thus generic technology is at the applied end of a spectrum of useful, but generally non-proprietary knowledge that begins with basic research but extends to tools, methods, materials, and basic processes.

Pre-competitive technology is defined the Department of Commerce as "research and development activities up to the stage where technical uncertainties are sufficiently reduced to permit preliminary assessment of commercial potential and prior to development of application-specific commercial prototypes."[28] (There is an implicit assumption that when consortia of competing firms are eager to collaborate in a government project, while fully aware of anti-trust limitations, the project is pre-competitive.)

Neither of these descriptions is operationally defined so that they can be applied directly to program selection; both serve as a general indication that the projects should not be product-specific and that funding them will not be anti-competitive. Rather, project results should diffuse broadly through a significant sector of the economy, thus "enabling" a more productive, and no less competitive, economy.[29] These categorical descriptors, however, do not prescribe the process used by ATP for the selection of projects from among those that pass the rather general initial screening for qualification as "pre-competitive generic." Scholars examining the projects actually funded in the first two years have questioned whether all of them were, in fact, "generic."[30]

The ATP has specified five standards by which proposals will be judged. They are: 1) the scientific and technical merits of the proposal; 2) potential broad-based benefits of the proposal; 3) technology transfer benefits of the proposal; 4) experience and qualifications of the proposing organization; and 5) the proposer's level of commitment and organizational structure.[31] Note that none of these filters selects specifically for "pre-competitiveness," and only criteria two and three hint at "generic-ness." Proposals are reviewed by ATP's panel of experts, and ranked on each point. While it is possible to quibble with the criteria themselves—the lack of sectoral focus was discussed in Chapter 2—the criteria are clear and open. The use of a panel of experts is analogous to the peer review system that works so well in basic science. The important attribute of the support of infrastructural research of value to industry is continuity and strong professional relationships between the basic and applied scientists doing the research and the engineers in industry who depend on it. Thus although the statute is not explicit, the ATP program has set up standards that are likely to be successful.

Pathbreaking technology programs to exploit emerging scientific breakthroughs that have industrial potential have been a hallmark of defense, energy, and space research. In each area the agency's development program has been supported by an associated, broadly-based scientific and engineering research activity to create the knowledge infrastructure on which the technological opportunity can be built. Given the fluid nature of leading-edge science and technology, all pathbreaking technology programs should stress flexibility as much as possible. The agency best known for its flexibility is DARPA (now ARPA). Project selection is left to the judgment of a program manager, who awards contracts to any organization that he or she deems appropriate.[32] The success of DARPA has often been attributed to the flexibility and independence given to the program managers and to the fact that the program operates largely outside of the glare of the public spotlight.[33] The DARPA program is generally judged to have been effective. While its processes are neither open nor peer reviewed, its fairness is rarely challenged (even though it operates primarily through sole-source selection), since DARPA is an agency of the customer for its developments: the Department of Defense.

The most successful selection process for the management of pathbreaking technology is probably the NIH program in support of biotechnology. In this case, NIH never had an explicit mission to create an industry. This may be a partial explanation for its success, since it didn't have to contend with the expectation that it would create jobs. Its basic research was tied in rather general ways to a clear social objective: finding diagnoses and treatments for the maladies encompassed by the names of the NIH institutes. The result was that NIH did exactly the right things: It generously funded the basic science that drove the vision of opportunity. It funded the tools and materials (i.e., the infrastructural technology) that allowed reduction to practice of the vision.[34] And it funded the clinical research that helped create the market. No other pathbreaking technology program comes close to this successful example.

There are many other examples of defense contributions to pathbreaking technologies that created industries. The four key attributes, present in the biotechnology case, are also present in the successful example of the laser industry: the basic scientific understanding, the infrastructure of tools and materials and process knowledge, the priming of a variety of applications markets, and the requisite patience and flexibility.

The operation of DARPA also shows some of the limitations that politics places upon would-be program models. The call for a "civilian DARPA" was based on DARPA's perceived success, which was agreed to be a function of its flexibility, independence, and anonymity. Curiously, by drawing attention to DARPA, its admirers endanger its anonymity. A civilian counterpart might not have the same flexibility and independence. It is unclear whether the American public would accept a program like DARPA in the civilian sphere, where managers would be given as much responsibility and authority, but their decisions would not be cloaked by the shield of national defense.

Along with the characteristics of clear selection criteria and program flexibility, these new technology development programs should have clear criteria for termination. Criteria for judging the ultimate success or failure should be articulated clearly from the beginning. For most pathbreaking, strategic, or high-entry-barrier projects, success means turning the technology over to the private sector and removing government support (or a shift in support to basic and infrastructural research). One of the great fears of opponents of efforts to assist technology commercialization is that the projects will become hard to kill because of the development of strong political constituencies. Cohen and Noll point to this as one of the factors in the failure of past U.S. government efforts in large-scale commercialization projects.[35] While this conflicts somewhat with the desire to focus research on long-term technical objectives, providing a credible way to kill a project is key to garnering legitimacy. (There is a third option besides termination and continuation for a development project that is not meeting its goals: to downgrade it from a likely source of commercial technology to more of a long-term, basic research project. The impending decision on the U.S. atomic fusion research program might be an example of this process.)

Armed with a set of criteria for defending (or challenging) proposals for supply-side policy in science and technology, and a set of program mechanisms for carrying them out, are the capabilities of any of the existing agencies adequate to this task? If not, what institutional forms should be adopted?

Institutional Choices for Civilian Technology Programs

Two sets of institutional choices must be made: First, what civilian or military agencies—new or existing—should be given responsibility for managing the civilian industrial R&D programs? Second, what institutions will be relied on for the performance of the civilian R&D—firms or groups of firms, national laboratories, universities, or combinations of them?

The agencies most often discussed as a lead agency for the role of sponsor of civilian industrial technology are:

•ARPA, formerly DARPA, given a broadened mission in dual-use technology. It was renamed in 1993 to signify the importance of its dual-use technology, and was also endowed with additional resources, to signify the importance of its dual-use mission.

•The National Institute for Standards and Technology (NIST) in the Department of Commerce, with ninety years' experience working with commercial industry on infrastructural technology.

•The Department of Energy, with great technical resources (and a $6.6 billion R&D budget) in its national laboratories, including about $3 billion in three nuclear weapons laboratories whose nuclear missions have shrunk dramatically.

•The National Science Foundation, devoted primarily to the support of academic science and engineering, but supporting a broad variety of activities to enhance the collaboration of university science and engineering with peers in industry.

•The National Institutes of Health, funding some $6 billion of biomedical R&D, which already plays the lead role in the promotion of the biotechnology industry, but is specialized to the general area of health and life sciences.

•NASA, with a large array of government laboratories in support of space missions, and several devoted to infrastructural technology in support of the aerospace industry (a rather specialized capability).

As discussed in Chapter 4, Congress focused most of its legislative energies during the Reagan and Bush administrations on enhanced technology transfer from these government-mission defined agencies to industry, relying on an outmoded spinoff paradigm. Critics of spinoff, even when it is assumed to be automatic and free, call attention to the mismatch between the institutional structure just described and the goal of energizing the technological competitiveness of a broad range of industries, going beyond defense, commercial aerospace, and health-related industries. Enthusiasts for a more aggressive government role have proposed the creation of new agencies as candidates for the competitiveness mission. Their hope is to couple the level of resources and independent judgment of military agencies with the commercial and political sophistication that will be required to build partnerships with a private industry that is generally skeptical of government efforts to help it. Two of these suggestions are:

•The "Civilian (D)ARPA," an agency to be placed in the Commerce Department, but with technology investment strategies fashioned after DARPA in the Defense Department.

•A Civilian Technology Corporation, proposed by the National Academies of Science and Engineering, intended to operate at arm's length from political and bureaucratic constraints.

We discuss the "civilian DARPA" first, then the existing agencies, including ARPA and NIST, the universities, the national laboratories, and then come back to the idea of the Civilian Technology Corporation, perhaps the most radical departure from tradition that is on the table.

A New Civilian DARPA?

On November 21, 1989, Senator John Glenn, chairman of the Senate Governmental Affairs Committee, introduced the "Trade and Technology Promotion Act of 1989" (S. 1978). The purpose of this bill was to create in the Department of Commerce an "Advanced Civilian Technology Agency" (ACTA) to support technologies for commercial industry by analogy to the Defense Department's Advanced Research Projects Agency (DARPA).[36] ACTA would create a revolving fund to support cost-shared public-private partnerships. The proposal, which was not enacted, is one of many that seek to enlist DARPA in the government's competitiveness strategy or clone it as a civilian DARPA (called CARPA) in the Commerce Department.

There is no reason, however, to believe that calling a new civil agency ACTA or CARPA would earn it a $2 billion budget, or that managers of a civilian agency could ever enjoy the freedom of action of DARPA's managers, who are cloaked in the autonomy that national security uniquely imparts. Many conservatives question whether a government agency can do a better job than a sector-specific venture capital firm at making technology investments in a commercial market.[37]

Another proposal, from the Carnegie Commission on Science, Technology and Government, suggested that emphasis be given to the dual-use technology mission of DARPA, changing its name to "NARPA": the National Advanced Research Projects Agency in recognition of its dual use activities, and giving it new authority to combine efforts with civilian agencies investing in the same technologies.[38] In the FY1993 defense appropriation, Senator Bingaman found a more acceptable name: reversion to the agency's original name, ARPA. A strong case for chartering ARPA with an explicit dual-use technology development mission is made by Jeff Bingaman and Bobby R. Inman.[39] This recommendation was given effect in the FY1993 Defense Appropriations Act.

ARPA

The fascination with DARPA as a model for an agency to make investments in the civil industrial technology base has four roots.

•DARPA enjoys an R&D budget of about $2 billion, more than 30 times the size of the Commerce Department's Advanced Technology Program.[40] The selection of a defense agency to lead the switch from defense to civilian funding of R&D is sometimes justified by the observation that it is easier to change the mission of an agency with a strong constituency that it is to transfer its funding to another agency that might otherwise be more appropriate.

•DARPA's role has been to invest in technologies of military potential far in advance of established service requirements; it enjoys a high degree of latitude in deciding what these technologies will be. It has enjoyed a reputation as a bridge between innovative ideas from the academic community to the proof of principle of new defense technologies.[41]

•DARPA has a small, technically expert staff, with a reputation for low overhead and quick decisions.[42] (Contracting is managed in part through the military services.) It enjoys a high degree of programmatic autonomy and a minimum of red tape.

•Since DARPA explores technologies not yet within the embrace of the military services and its defense industry, it has experience in dual-use technologies, such as high-performance computing and sophisticated materials. It thus has direct experience with a number of commercial technologies.

As documented in *Beyond Spinoff*, the defense industry is too isolated from leading-edge commercial technology, which is often years ahead of the technology in fielded weapons systems, to ensure that our military maintains a clear technical lead. For that reason ARPA, under its new (and original) name, is being given an explicit mandate to develop dual-use technology in response to anticipated defense needs. It should be encouraged to enter into Cooperative Agreements with commercial firms (and with civil agencies) to develop technologies of value to both civil and military uses.

Title IV of the FY93 Defense Appropriations Bill authorized changing the name of DARPA back to ARPA by way of recognition of this dual-use technology role. It provides almost half a billion dollars in new funds for dual-use technology, training, and industrial extension services to ease the impact of defense budget reductions, and to pump new resources into commercial firms.

ARPA can be especially effective in support of pathbreaking technologies with potential to provide qualitative advantage to U.S. military forces while opening up the possibility of new industries. Collaboration with civil agencies, especially the Department of Commerce, can help ensure that the voice of industry is properly reflected in such activities.

Title IV is to be implemented through a Defense Technology Conversion Council (DTCC), with representation from ARPA, NIST, DOE (the weapons labs), NSF, and NASA. Its mission is:

To stimulate the transition to a growing, integrated, national industrial capability which provides the most advanced affordable military systems and the most competitive commercial products....Execution of the TRP programs will be done on a distributed basis, with oversight by ARPA, and with execution by the Military Departments and DoD agencies, NIST, DOE, NSF, and NASA.[43]

With the collaboration of these other agencies, ARPA will manage the Technology Reinvestment Project (TRP), to ease the transition from military to civilian business for industry, to promote the commercialization of military technologies (spinoff) and to give Defense services access to commercial technologies ("spin-on"). One of the more remarkable and complex features of this program is the establishment of a "one-stop shopping" mechanism for applicants. A toll-free telephone number (1-800-DUAL-USE) has been established for inquiries. Firms applying for support are guided through a matrix of types of technical activities and programs in the participating agencies.

This mechanism for associating five departments and agencies in the management of $471 million of defense appropriations is a response to the fact that defense conversion and the support and commercialization of dual-use technologies involve both military and civil activities. The program goals thus overlap those of the agencies. For example, the TRP provides ARPA with $87.4 million for manufacturing extension programs, an activity assigned to NIST by the 1988 Omnibus Trade and Competitiveness Act. NIST's appropriation for its activity is much smaller, but it has experience that ARPA does not have with small manufactures and the state-based programs set up to help them.

At this writing (May 1993) it is too early to see if the organization of the Defense Technology Conversion Council will become a significant institutional innovation, or whether it will join the ranks of dozens of other inter-agency coordinating committees that have minimal influence. What can be concluded is that the administration, in the pragmatic style Americans seem to prefer, has not made a choice between the use of military versus civil agencies for the leadership role in civilian industrial technology policy. In FY93, the defense budget contains the great majority of new funds for such activities. But the Clinton strategy calls for a shift of $8 billion of this money from military to civil investment, and identifies NIST in the Commerce Department to receive $117 million in new funding in an FY93 supplemental appropriation for its work in this area. It is therefore unclear what the agency roles will be in future years. At present, ARPA is in the dominant position because of its access to most of the new resources.

NIST and ATP

The Advanced Technology Program, managed by the National Institute for Standards and Technology (NIST) in the U.S. Department of Commerce, provides cost-shared funding to commercial firms, or consortia of firms, in support of commercially relevant technology. The budget request for this program in FY93 is $69 million, but the Clinton administration proposed to add a supplemental appropriation, and to expand NIST's activity substantially over the next four years. As currently administered, the ATP program does not give priority to any industry or technology sector. The program is not tied to a list of "critical technologies," although NIST provided such a list in 1990. (See Chapter 2.) Instead, the program is open to unsolicited proposals from firms for projects of their own selection.

Proposals are subject to a review process that first scores the projects for technical excellence. Those found excellent—creative and likely to advance the state of the art—are reviewed for their business promise and the capabilities and commitment of the proposing firms. Thus the selection process identifies projects that do not *a priori* follow any list of pre-determined critical technologies. They are not grouped in any way by industry, or by a generic problem they may propose to attack. This determination not to define technical or business areas in which proposals are invited (as the interagency Technology Reinvestment Project does), serves to protect against accusations of "picking winners," but makes it exceedingly difficult to detect the economic effect of the projects through subsequent sectoral analysis.

With three years' experience (at very modest levels of funding), the ATP program has moved steadily toward the generic end of the program scope, focusing increasingly on process and production technologies that can be used in a variety of industries and give promise of increasing productivity and quality. Given the many decades of NIST experience in its own laboratories with process and materials characterization and test methods, this seems a fruitful direction for the focus of ATP.

Congress also assigned NIST the mission to establish, in collaboration with the states, a series of Manufacturing Technology Centers and to help the states strengthen their industrial extension services through both the MTCs and a new series of 100 Manufacturing Outreach Centers. (See Chapter 6.) Over the years NIST has hosted over 1000 resident "guest workers" from industry, who collaborate with NIST scientists on infrastructural research of generic industrial importance. This collection of activities makes NIST, which forms the cornerstone of the Commerce Department's Technology Administration, the logical focus of "lead agency" responsibility for the federal government's attempts to help the technological performance of U.S. industry. The Technology Administration includes NIST, the National Technical Information Service (NTIS), and the National Telecommunications and Information Agency (NTIA), and is headed by the Undersecretary for Technology. However, the organizational structure of NIST and of the Technology Administration is inadequate to support the new missions and rapid growth of program activity anticipated by the administration. We suggest what might be done below.

The Universities and the National Laboratories

NIST's ATP program and the ARPA dual-use technology program avoid a serious barrier to their effectiveness by avoiding dependence on transferring the technology from the federal R&D performer to the firms hoping to commercialize that R&D; instead these programs fund the work in industry directly. The universities and the national laboratories (primarily those operated by DOE, NIH, and NASA) perform their R&D in their own facilities and transfer the technology to industry through a variety of mechanisms. The most efficient of these mechanisms is enjoyed by the universities: it is the career path of their science and engineering graduates and research fellows into industry. By that device the universities transfer both codified knowledge and "embedded" or tacit knowledge. Students' training, especially at the doctoral level, acquaints them with the contemporary state of knowledge.

The Universities
The universities, beholden to both federal and state governments for financial support, have tended to exaggerate the extent to which they offer the solution to the nation's competitiveness problems.[44] There are rather severe limits to the extent that universities can engage in commercially related activities, and many are testing those limits today. However, the extent of linkages between university research and industry is impressive indeed. A study for the Ford Foundation has searched out all the University-Industry Research Centers, and found 1050 of them, spending $4.5 billion annually.[45] This funding comes in almost equal parts from federal agencies, industry, and other sources that include the universities and the states. The distribution of support across federal agencies is surprisingly uniform and reflects broad-based support.[46]

The industry coverage is quite broad too, spanning chemicals, pharmaceuticals, computers, software, electronics, petroleum, and coal. This study strongly suggests that the universities are more deeply engaged in relationships with industry than has commonly been thought. The greatest value of this level of university involvement with industry is access by the university's science, engineering, and management students to industrial culture and problems, and the access that participating firms have to students whom they can recruit into their employ. Through the students, universities represent a uniquely effective mechanism for technology diffusion. The National Science Board, the policy-making body operating the National Science Foundation, has given serious attention to the Foundation's role in the new competitiveness policy. (How the universities and government will work out their new relationship, with industry as an actively participating third party, is the subject of Chapter 7.)

The National Laboratories

The government's directly operated and contractor-operated national laboratories perform over $21 billion in R&D annually. Many of these laboratories are very sophisticated and are well experienced at cross-disciplinary teamwork bridging science and engineering. Given the massive shift in national priorities from defense to the domestic economy, many of these laboratories, especially the three DOE nuclear weapons laboratories, are eager to participate in the new civilian technology missions. Unlike the ARPA dual-use technology program and the NIST ATP program, which contract directly with commercial firms, the national laboratories would spend the money in-house, although increasingly in collaboration with firms. The weapons labs represent an institutional form inappropriate to the needs of a strategy aimed at enhanced commercial industry performance, but given the high levels of capability they possess and the political pressure to sustain them at this level, they are attractive to the administration and Congress as a vehicle for contributing to the commercial technology base. Their role is the subject of Chapter 4.

The NIH laboratories are a different story. They engage primarily in basic research, but this work has been highly productive in stimulating new firm creation in biotechnology and in sustaining the rapid progress of the industry. Indeed, the success of this linkage has led some members of Congress to insist that NIH use its leverage under its cooperative research agreements with pharmaceutical firms to contain the growth of drug pricing.

The NASA aeronautical laboratories (those that formed the core of the National Advisory Committee on Aeronautics, from which NASA was formed) continue to support infrastructural research of potential value to the air transport industry. The NACA experience between the two World Wars, generally credited with a major contribution to the rapid maturity of the U.S. aircraft industry, sets a model for the contribution of government laboratories to a young industry. It is less clear that the laboratories are as influential today, with the maturity of the industry and its winnowing down to two very large firms.

The Civilian Technology Corporation

In 1992, a highly experienced Panel on the Government Role in Civilian Technology, assembled by the Committee on Science, Engineering and Public Policy of the National Academy of Science and the National Academy of Engineering, evaluated the existing agencies as candidates for the lead role in government stimulation of commercial technological performance. They concluded that a military agency was not appropriate, and found the civilian agencies too vulnerable to political interference to have a good chance at success.[47] They recommended that a Civilian Technology Corporation (CTC) be chartered by Congress, and governed by a board appointed by the president and confirmed by the Senate. It would receive a one-time appropriation of $5 billion in public funds, and would have the authority to make venture capital investments as well as R&D grants and contracts, with recapture of profits if it chose. This "arm's length" relationship to government would, advocates say, increase the likelihood that bold and non-political choices of technology investments would be made.

Critics point to the suggestion that the board of directors of the CTC, not the president, appoint its CEO. They recall that the reason President Truman vetoed the original bill to create the NSF (then called the National Research Foundation) was his insistence that, if public funds were to be expended, the CEO must be accountable to him, just as he was accountable to Congress and the people for the use of their tax monies. Truman had his way in the present form of the National Science Foundation, which is managed by a board appointed on fixed and staggered six-year terms, and a director who serves at the pleasure of the president, not the board. The political pressure for strict accountability is even stronger today than it was then. Asking Congress for a check for $5 billion up front, without direct oversight by Congress and control by the president of the Corporation's activities, appears unrealistic.

Conclusions

The leading role of ARPA in the multi-agency Technology Reinvestment Project (TRP) exemplifies the American propensity for
pragmatism in government. President Clinton announced his
intention to shift some $8 billion from defense to civil agencies, but
because this requires a full Congressional budget cycle to accomplish, it was quicker to assign ARPA the authority to undertake a
number of the civilian technology activities under the heading of
defense conversion. ARPA's leadership of the Technology Reinvestment Project may evolve into the "national advanced research
projects agency" envisioned by the Carnegie Commission study,
acquiring authority to invest in both military and civil technologies.
On the other hand, the stringencies of military R&D budgets, and
industrial concerns about the influence of "defense industrial
culture" on ARPA's commercially oriented activity, may result in
the current ARPA role being temporary, with a reversion to its focus
on pathbreaking (defense-unique and dual-use) technologies in
support of a leaner, more focused defense technology effort. It is
unclear in the spring of 1993 which of these two outcomes will
result. The best outcome would be a transitional arrangement in
which ARPA and the Commerce Department collaborate to build
the kinds of capabilities that ARPA will need to acquire commercial
technologies for military use and that the Commerce Department
will need to assume the leadership role in civil technology that it
logically should have.

For NIST and the Technology Administration of the Department
of Commerce to assume the role of lead agency in defining how
government-funded R&D can best contribute to industrial competitiveness, weaknesses perceived to handicap the Department
must be addressed. NIST enjoys strong support from technical
executives in industry and the respect of the U.S. scientific community. However, the Commerce Department receives ambivalent
political support from private industry.[48] For the civilian technology initiatives of the Clinton administration to be successful, they
must receive whole-hearted support from the business community.
The Commerce Department is more likely to be successful at
generating that support than any other department, but it must be

staffed with experienced industrial technology executives who can manage the department's activities to win that support.

NIST is one of the nation's most valuable research laboratories, with unique ties to industry. It is a good place to incubate new technology activities such as those authorized in the 1988 law. But as these activities mature and grow, they should not be allowed to dwarf the research laboratory which sustains NIST's core competence. It is unreasonable to expect a fine, conservatively managed scientific laboratory like NIST to manage the new missions (ATP, Manufacturing Technology, and Industrial Extension) with all their political and managerial complexities.

Instead, the new activities should be organized in a new agency, which might be called the Agency for Industrial Technology, built on a core staff from the current NIST staff, and perhaps located in facilities on the NIST grounds in Gaithersburg, to have easy access to experts on the NIST staff. This new agency would become the most politically visible, and probably the most rapidly growing agency in the Technology Administration. Its responsibility for a major acceleration of industrial extension services would require political sophistication to deal with the fifty state governments. (See Chapter 6.) The new agency would have to play a strong role in the national data network facilities and services through which private as well as public sources of business and manufacturing technology information will flow. (See Chapter 5.) It would also manage the Advanced Technology Program (ATP), which requires that it address the criteria for project strategy and selection discussed in this chapter. It must also coordinate related activities in other agencies having much bigger budgets and more political "clout."

The Advanced Technology Program should be recast to focus on a smaller number of areas of technology of greatest promise, selected with the advice of industrial and academic experts. The criteria for selection should be made much more explicit and reflect sound economic criteria for public investment. As noted, two key areas for investment are pathbreaking technologies, where the technical risks and time to commercialize are too long for adequate private investment, and infrastructural research of broad application to productivity growth but low appropriability to specific firms. The areas of research focus in the NIST in-house research program are an excellent model for the kind of work the ATP program should be supporting. Critical technologies lists would be largely irrelevant in that event.

The proposed new Agency for Industrial Technology should also be prepared to receive new assignments in technology monitoring and competitive assessment (or "benchmarking science and technology" in the language of the White House). This mission to compare the standing of American industrial technology with the best commercial technology internationally is a major feature of H.R. 820, The National Competitiveness Act of 1993, and a companion bill in the Senate, S. 4, which would establish in the Department of Commerce an Office of Technology Monitoring and Assessment. Assessing competitiveness has been widely recommended as a prerequisite for selecting the areas of technology deserving of federal R&D investment.[49] An additional role for the new agency is in the management of international technological cooperation. In the past such projects have been commonplace in specialized agencies such as defense, NASA, and Energy. More recently, the Japanese have proposed a number of multi-national projects with an industrial focus, such as the Intelligent Manufacturing System project and New Earth 21, a 100-year-long environmental project aimed at industrial ecology issues.

Such an Industrial Technology Agency, together with NIST, the National Technical Information Service, and the National Telecommunications and Information Agency, would be the core of the Technology Administration. Together with the proposed Civilian Technology Development Advisory Committee, advising the Undersecretary for Technology on matters of policy, planning, execution, and evaluation of the Civilian Technology Program proposed in H.R. 820 and S. 4, the Technology Administration could hope to gain the legitimacy and experience it needs to lead the federal effort in civilian technology.[50]

There are widespread variations in public and business confidence that the federal government could implement these policies with vigor and effectiveness, avoiding the many pitfalls described in this book.[51] The Clinton administration expresses a new confidence in the ability of government to collaborate with the private sector and the state governments to address long-term opportunities to enhance U.S. economic performance. Skeptics do not believe government agencies can do this job competently, and even if they could, would quickly fall victim to political distortion and capture by program beneficiaries. An effort to prove the skeptics wrong is long overdue.

Notes

1. Maurice Stans, Hearing of the House of Representatives Committee on Science and Astronautics on "Science, Technology, and the Economy" (Washington, D.C.: U.S. Government Printing Office [U.S. GPO], July 27–29, 1971), p. 17.

2. William Clinton and Albert Gore, Jr., *Technology for America's Future*, February 22, 1993, p. 1.

3. OECD, *Main Science and Technology Indicators*, 1990–91 (Paris: Organization for Economic Cooperation and Development [OECD], 1990) p. 16, Table 2, and p. 20, Table 14.

4. Clinton and Gore, *Technology for America's Future*.

5. John A. Alic, Lewis M. Branscomb, Harvey Brooks, Ashton B. Carter, and Gerald L. Epstein, *Beyond Spinoff: Military and Commercial Technologies in a Changing World* (Boston: Harvard Business School Press, 1992), p. 89.

6. Clinton and Gore, *Technology for America's Future*, p. 33. The Carter administration launched its own "reinvent the automobile" project in its last year, only to see it terminated by the Reagan administration. Known as the Civilian Automotive Research Program (CARP), it was also criticized as representing a political convergence of three interest groups but without a common goal. See L.M. Branscomb, "Opportunities for Cooperation between Government, Industry, and the University," *Annals of the New York Academy of Sciences*, Vol. 334 (December 14, 1979), pp. 221–227.

7. Public Law 97-219.

8. *The SBIR Basics: Questions and Answers* (Phoenix: Project SBIR West, March 8, 1993), p.1. SBIR West is a non-profit association encouraging firms in Iowa and adjacent states to take advantage of the SBIR program.

9. The 1992 amendments, discussed in the next paragraph, do broaden the scope of the act to authorize investment in the critical technologies mentioned in the DoD and OSTP "lists." It is unclear whether this will lead agencies to stray further from their core missions than they already have in the SBIR program.

10. The Kennedy administration's Civilian Industrial Technology Program is an example, and is discussed briefly in Chapter 2.

11. *Public Papers of the Presidents of the United States: Jimmy Carter 1979*, Book II, June 23–December 31, 1979 (Washington, D.C.: U.S. GPO, 1980), pp. 2068–2974.

12. The pathbreaking and infrastructural technologies criteria were proposed and discussed in *Beyond Spinoff.* In both cases, technical experts from industry must have a major voice in allocating resources, and firms should also be partners with government and universities in the conduct of the work. Alic, et al., *Beyond Spinoff,* chap. 12.

13. Linda Cohen and Roger Noll, *Technology Pork Barrel* (Washington, D.C.: The Brookings Institution, 1990).

14. Thomas J. Allen, *Managing the Flow of Technology* (Cambridge, Mass.: MIT Press, 1984).

15. Roy Rothwell, "Developments Towards the Fifth Generation Model of Innovation," *Technology Management and Strategic Management,* Vol. 1, 1992, pp. 73–75.

16. George Heaton, "Commercial Technology Development: A New Paradigm of Public-Private Cooperation," *Business in the Contemporary World,* Autumn, 1989, pp. 87–98.

17. Ibid., p. 92.

18. Eric von Hippel, "Cooperation Between Rivals: Informal Know-How Trading," *Research Policy,* Vol. 16 (1987), pp. 291–302. General Accounting Office, *Diffusing Innovations: Implementing the Technology Transfer Act of 1986* (Washington, D.C.: U. S. GPO, 1991). Jon Soderstrom, Emily Copenhauer, Marilyn A. Brown, and John Sorenson, "Improving Technological Innovation through Laboratory/Industry Cooperative R&D," *Policy Studies Review,* Vol. 5, No. 1 (August 1985), pp. 133–144.

19. This critique is made in a general way in Harvey Brooks' comments on technological monocultures in "Typology of Surprises in Technology, Institutions, and Development," in William Clark and R.E. Munn, eds., *Sustainable Development in the Biosphere* (New York: Cambridge University Press, 1986), pp. 325–350. It is made in specific reference to technological innovation in James M. Utterback's "Invasion of a Stable Business by Radical Innovation," in Paul Kleindorfer, ed., *Management of Productivity in Technology in Manufacturing* (New York: Plenum Press, 1986).

20. David Mowery, "Economic Theory and Government Technology Policy," *Policy Sciences,* Vol. 16 (1983), pp. 27–43; W.H. Lambright and Albert Teich, "Federal Laboratories and Technology Transfer: An Interorganizational Perspective," in D.E. Cunningham, et al., eds., *Technological Innovation* (Boulder, Colo.: Westview Press, 1977), pp. 425–440.

21. Clinton and Gore, *Technology for America's Economic Growth,* p. 8.

22. The idea that local industrial clusters are a source of competitive strength appears in Michael Porter, *Competitive Advantage of Nations* (New York: Free Press, 1990); and in Michael Piore and Charles Sabel, *The Second Industrial Divide* (New York: Basic Books, 1984).

23. Charles Sabel, "Studied Trust: Building New Forms of Cooperation in a Volatile Economy," in Frank S. Pyke and Werner Sengenberger, eds., *Industrial*

Districts and Local Economic Regeneration (Geneva: International Institute for Labor Studies, 1990); Philip Shapira, *Modernizing Manufacturing* (Washington, D.C.: Economic Policy Institute, 1990).

24. Thomas E. Seidel, Chief Technical Officer, SemaTech, MIT lecture, March 17, 1992.

25. Public Law 98-462.

26. Clinton and Gore, *Technology for America's Economic Growth*, p. 10.

27. Title 15 of the Code of Federal Regulations, section 295.2 (b), under 15 U.S.C. 271 et seq., and section 5131 of the Omnibus Trade and Competitiveness Act of 1988 (P.L. 100-418).

28. Ibid., section 295.2 (g).

29. In several speeches by President Bush beginning in April 1990, and in D. Allan Bromley's *U.S. Technology Policy*, issued by the White House in September 1990, the administration added the adjective "enabling" to the "pre-competitive generic" qualifiers set up by the 1988 Omnibus Trade and Competitiveness Act to delimit the federal role in support of commercial technology.

30. Vincent J. Ruddy, "Criteria and Processes to Support Generic, Pre-competitive, and Enabling Technology Development," Kennedy School of Government Policy Analysis Exercise, April 9, 1991.

31. National Institute for Standards and Technology, *ATP Proposal Preparation Kit* (Washington, D.C.: U.S. Department of Commerce, 1992).

32. Burton Edelson and Robert Stern, *The Operation of DARPA and its Utility as a Model for a Civilian ARPA* (Washington, D.C.: The Johns Hopkins Foreign Policy Institute, November 1989).

33. Lois Ember, "Major Change in Offing for DARPA, Key Defense Research Agency," *Chemical and Engineering News*, September 14, 1992.

34. It should also be noted that the FDA regulations that create barriers to market entry for pharmaceuticals also force the development of much of the infrastructural technology, which has to be developed to win approval to develop and market a drug, although few would conclude that a substantial economic price has not been paid for the medical efficacy and public safety the regulations are there to serve.

35. Cohen and Noll, *The Technology Pork Barrel*.

36. *Managing Critical Technologies: What Should the Federal Role Be?* Research Report No. 943 (Washington, D.C.: The Conference Board, December 14, 1989), p. 4.

37. George Donohue, Richard H. Buenneke, Jr., and Wayne G. Walker, "Why Not a Civil DARPA?" *Rand Issue Paper*, Vol. 1, No. 1, November 1992.

38. *Technology and Economic Performance* (New York: Carnegie Commission on Science and Technology, 1991).

39. Jeff Bingaman and Bobby R. Inman, "Broadening Horizons for Defense R&D," *Issues in Science and Technology*, Vol. 9, No. 1 (Fall 1992), pp. 80–85.

40. In FY92 DARPA's budget of $1,586.3 million comprised: Research, $115.8 million; exploratory development, $744.4 million; advanced technology development, $657.6 million; and management and support, $68.5 million. The FY93 budget adds $473 million for dual-use technology and defense technology conversion to civil economic activity, all under a interagency program called the Technology Reinvestment Initiative.

41. DARPA has been criticized in the past, however, for failure to win adoption of the technologies it promotes into the weapons systems of the military services.

42. FY92 personnel: 142 civilians, 25 military, 16 IPAs, total of 183. But two-thirds of DARPA's expenditures are contracted by service and other agencies for DARPA, which minimizes DARPA's administrative burden.

43. ARPA, *Program Information Package for Defense Technology Conversion, Reinvestment, and Transition Assistance* (Washington, D.C.: Advanced Research Projects Agency, Department of Defense, March 10, 1993), p. 2-1.

44. Harvey Brooks, "University-Industry Cooperation as Industrial Strategy," in S. B. Lundstedt and Thomas H. Moss, eds., *Managing Innovation and Change* (Dordrecht, Netherlands: Kluwer Academic Publishers, 1989), chap. 4, pp. 35–46.

45. Wesley Cohen, Richard Florida, and W. Richard Goe, *University-Industry Research Centers in the United States*, Center for Economic Development, Carnegie Mellon University, A Report to the Ford Foundation, June 1992, to be published in book form in 1993.

46. NSF provides 54.9% of UIRC's aggregate support; Defense, 45.1%; Energy, 33.9%; NASA, 27.3%; and NIH, 26.7%. These percentages add to 187.9%. The centers were asked from what agencies they receive support. The data show that on average each center is supported by at least two agencies. The figures do not suggest that all agencies are investing in centers at comparable levels. NSF, for example, may support more than half of the centers, but with very modest grants.

47. Committee on Science Engineering and Public Policy, *The Government Role in Civilian Technology: Building a New Alliance* (Washington, D.C.: The National Academy Press, April 1992).

48. Dr. Lawrence Kushner, then Deputy Director of the National Bureau of Standards (now NIST), once described the Department of Commerce as the "only department of government with a hostile constituency" (private conversation with Lewis Branscomb, then Director of NBS).

49. See Alic, et al., *Beyond Spinoff*, pp. 400–401.

50. 820, Title III, subtitle D.

51. On September 14, 1992 a workshop on Federal Roles in Commercial Technology at the Kennedy School of Government, Harvard University, explored the subject of federal roles in commercial technology. There was little difference of opinion among the participants on desirable technology policies, but extensive differences over the likelihood of competent and politics-free execution.

4

National Laboratories: The Search for New Missions and New Structures

Lewis M. Branscomb

The end of the Cold War has left the DOE weapons labs scrambling to define new missions for themselves, yet they are all reaching for the same new missions. The Task Force . . . skirts a fundamental question that must be addressed: With the end of the Cold War, do we still need three nuclear weapons labs, each funded at approximately one billion dollars per year and each with employment of about 8,000 people? It seems to me . . . that the . . . answer is "no."

The Hon. George E. Brown, Jr., Letter to Hon. James D. Watkins, February 9, 1992.

The science and technology base of the Laboratories provides what I call the infrastructure for solving large problems of great complexity. It is this infrastructure that I propose to bring to bear on the question of the competitiveness of our industries and businesses. This should be done in partnership with business and universities. . . . Business can provide the market pull on the talents of the Laboratories that will assure their work is relevant.

The Hon. James D. Watkins, Letter to Hon. George E. Brown, Jr., February 18, 1992.

Over the past forty years the United States built a system of government-funded research and engineering laboratories that conducts thirteen percent of the nation's research and development, exceeding the total R&D performed in either Britain or France. With Cold War's end, and the shift of the nation's priorities toward economic competitiveness, many of these 700 laboratories are scrambling for new missions.

These laboratories are very diverse in purpose, institutional structure, and capability. Some, such as the National Institute for Standards and Technology (NIST), were established to help commercial industry and are well suited to expand their traditional roles. Others have demonstrated that collaboration with industry can stimulate economic growth. Fundamental scientific research at the National Institutes of Health (NIH), for example, has stimulated a fast-growing biotechnology industry. Most of the laboratories, however, are dedicated to defense, energy, or space-related federal missions and confront both tight budgets and shifting priorities that raise questions about their future roles.

This chapter deals with the roles these laboratories can play in the nation's new technological priorities. Through a series of statutes passed since the early 1980s, Congress has been encouraging all of the laboratories to transfer their technology to industry, as a means of addressing industry's competitiveness problems. The primary mechanism is the Cooperative Research and Development Agreement (CRADA), authorized by Congress in 1986.[1] This chapter describes the diverse types of government-funded laboratories from the perspective of their usefulness in contributing to the economy. We focus first on the Department of Energy's nuclear weapons laboratories, returning to NIST and NIH at the chapter's end.

We explore a number of questions about the effectiveness of technology transfer efforts from defense and nuclear weapons laboratories, about the appropriateness of using public funds to subsidize activities with commercial firms, and about the structure of the laboratories themselves. The valuable national asset these laboratories represent can be put to very good use if three conditions are met: the laboratories are given more authority for self-management; they focus their activities on a limited number of core areas of technical competence; and they avoid the inappropriate use of public funds. Even then, however, the laboratories could find themselves in difficulty if public expectations for their role in the domestic economy are not realistic.

The three nuclear weapons laboratories of the Department of Energy represent a particularly conspicuous example of the challenge of responding to policy change, both because of their great size and capability, and because of the dramatic reductions in the nuclear weapons work that may be required of them in the future. The nuclear weapons laboratories were structured to pursue government development goals that called for isolation in a secure environment, rather than accessibility to commercial firms. Thus the federal government finds itself with an increasingly anachronistic national laboratory structure designed for Cold War development, facing a national demand to apply the laboratories to sustaining a highly diverse, market-driven, commercial industry. The laboratories are eager to respond in order to stabilize their futures, but they face a number of handicaps.

The Clinton administration proposes that the Department of Energy accelerate the technology transfer effort of its laboratories by authorizing the laboratories to spend up to ten percent of their appropriated funds on these activities.[2] Bills have been introduced into both the House and Senate to this effect. However, as we argued in Chapter 3, the government is not likely to be the primary source of commercially useful technology. Enhancing the competitiveness of private industry calls for a decentralized strategy that places more initiative in the hands of private industry, and focuses on better utilization of all existing technology, not limited to technologies developed in government-funded laboratories.

Expectations for increased contributions to the economy from government research spinoffs may have served to entrench Cold War institutional arrangements by broadening the constituency of the national laboratories.[3] By accepting the responsibilities inherent in the Federal Technology Transfer Act, the laboratories gave their Congressional delegations more powerful arguments against budget cuts and helped defuse the argument made by universities that they are more effective at technology transfer to industry by virtue of their educational activities.[4]

Thus the defense-related laboratories face, simultaneously, the threat of declining defense budgets and the call to transform their structures and cultures to adapt to new missions in industrial competitiveness, environmental technology, and other areas.

National and Other Federally Supported Laboratories: A Typology

Of FY 1992 federal expenditures of $70.4 billion spent on R&D, some $24,861 billion (over 30 percent of the total) is spent in the laboratories, and in their headquarters administration, which is entirely funded by federal agencies. For budgeting purposes, federally-funded R&D performers are divided into five groups (see Figure 4-1). Intramural laboratories are those owned and operated by government agencies.[5] They are staffed primarily by government employees, usually civil servants. Research contracted to profit-seeking businesses is reported in the table as "industry." The university category includes institutes and centers that are integrated into the life of a single university. Federally funded research and development centers (FFRDCs) are not-for-profit institutions operated for, but not by, government agencies to support the agencies' missions.[6] Contracts to operate FFRDCs may be held by universities, profit-seeking business firms, or not-for-profit corporations. Appendix 4-A shows the resources of FFRDCs by sponsoring agency and by the institutional form of the managing contractor. For example, the Los Alamos National Laboratory in New Mexico is operated for the Departments of Energy and Defense by the University of California; the Jet Propulsion Laboratory is operated for the National Aeronautics and Space Administration (NASA) by the California Institute of Technology.[7] The Oak Ridge National Laboratory is, by contrast, operated by Martin-Marietta Corporation, and the Savannah River Laboratory is operated by EG&G Corporation. Finally, some FFRDCs are contained within free-standing, not-for-profit corporations (such as MITRE Corporation and RAND Corporation). Still others, a number of which provide scientific facilities used by many universities and other laboratories, are operated by consortia of universities established for the purpose. Examples include the Brookhaven National Laboratory, operated for the Department of Energy by Associated Universities, Inc., and the Kitt Peak National Observatory, operated by the Associated Universities for Research in Astronomy.

Some eighteen of the FFRDC laboratories had R&D budgets of over $100 million in FY1990; the number is larger today. (Note that the total budgets of many of the labs, especially the weapons laboratories, are substantially higher because of work other than R&D.) These laboratories are listed in Appendix 4-B.

The Directly Operated Federal Laboratories

The total obligations for the directly operated, or intramural, federal laboratories in 1992 are estimated to be $17.645 billion,

Figure 4-1 Federal Obligations for R&D by performer, FY92 (estimated in $ millions)

Total	Intramural	Industry	Universities	FFRDC	Other
70,427	17,645	31,929	10,475	7,216	3,160

From National Science Foundation, *Federal Funds for Research and Development: Fiscal Years 1990, 1991, and 1992,* Volume XL, NSF 92-322 (Washington, D.C.: National Science Foundation, 1992), p. 51.

more than twice the $7.2 billion spent in FFRDCs. This figure includes the headquarters expense directly attributable to managing the laboratories. The federal agencies whose intramural (civil service) laboratories spent more than $100 million in FY 1992 are shown in Appendix 4-C.

Many of these laboratories have come under criticism over the years, because their missions are outdated, their effectiveness is compromised by bureaucratic government constraints, or they have been unable to attract the best scientific talent, most experienced management, or up-to-date equipment. The Department of Defense has embarked on an effort to identify ways of reducing its intramural R&D effort by up to 50 percent.

The first U.S. government laboratory was the U.S. Coast and Geodetic Survey, followed by the Geological Survey and the National Bureau of Standards (now NIST), all before the turn of the century. Other major civil service laboratories include the U.S. Public Health Service Laboratories (now NIH), the Naval Research Laboratory, the Langley, Ames, and Lewis Research Centers of NASA (the loci of much of the nation's aeronautical research work before World War II), and the Agricultural Research Service at Beltsville, Maryland, and elsewhere.

As government employees, researchers in intramural laboratories can perform technical work and make decisions as officers of the government. They can negotiate contracts for technology or equipment, make regulatory rulings, and take other official actions the contract personnel in an FFRDC cannot do. Most civil service scientists, however, only rarely act as agents of the government. They are reluctant to play any kind of regulatory role, preferring to devote themselves to research. Having taken an oath of office, civil servants are accountable to the president, Congress, and the public. Because the head count, mission scope, and funds for expenditure are specified by Congress annually, these institutions are under close political control. Their growth is unlikely to be rapid; their missions are stable to the point of inflexibility; their political freedom to act in their own interest is limited. The conventional wisdom that civil service laboratories are low in competence is an inaccurate generalization, however. Although there are examples enough to justify concern, particularly in the defense sector, government laboratories such as NIH, NIST, the U.S. Geological Survey, the Smithsonian Institution, and the Naval Research Laboratory have sustained world-class reputations for technical excellence and valued performance over many years. Their scientists have gathered Nobel Prizes and many other scientific honors. They pride themselves on their integrity and commitment to the public interest.

The intramural laboratories cannot be expanded rapidly or easily moved to new locations. They are very hard to shut down, since each has its own statutory existence, beholden to the committee of Congress responsible for its enabling statute. The best of these laboratories, such as NIST, NRL, and NIH, depend on decades-long traditions of high-quality work and have earned some special dispensations from government regulations concerning personnel matters. But most of them find it hard to shift the laboratory mission to new goals when the need arises, since their goals are usually defined in statutes. It is also difficult for most of them to compete with the contract laboratories in pay, flexibility of personnel management, capital facilities, and working conditions.

National Laboratories
The largest federally-financed research and development laboratories with the most diversified technical capabilities are called, informally, "national laboratories," although the designation "national laboratory" has no formal status. Most of the Department of Energy's multipurpose laboratories come to mind; a number of them, such as Los Alamos National Laboratory and Oak Ridge National Laboratory are in fact so named.[8] So, too, are some of the government's largest and best known intramural laboratories, such

as the National Institute for Standards and Technology (NIST) and the National Institutes of Health (NIH).

The government-owned, contractor-operated (GOCO) national laboratories (FFRDCs) evolved as a new institutional form during and after World War II. GOCOs were thought to have three advantages over civil service laboratories:

•Contract laboratories escape the strictures of civil-service pay schedules and government administrative regulations, and therefore should be more efficient and effective. This advantage has been realized in enough cases to validate the concept.

•A contract laboratory can be created through administrative action, if provision is made in the sponsoring agency's appropriation; it does not require establishment by statute. For the same reason, it should be easier to alter a laboratory's mission or even to close it down.

•The independence of the contract laboratory can be an advantage to its sponsoring agency, for it removes the performance of technical work one step from the focus of political pressures. Yet such a laboratory can fulfill most of the technical functions of its parent agency. Official decisions are made by the headquarters staff on advice from the contract laboratory.[9]

Throughout the growth of federal R&D activities after World War II, new R&D institutions were almost invariably created outside the civil service.[10] It has not been demonstrated, however, that the national laboratories are any easier to shrink, shut down, or redirect than are civil service laboratories. Vested interests in the laboratories' payroll and in subcontracts to the communities in which they are located are powerful political forces for stability in contract laboratory budgets and missions.

Department of Energy Weapons Laboratories
The Lawrence Livermore Laboratory in California, the Sandia Corporation, and Los Alamos National Laboratory in New Mexico conduct some $3 billion in R&D and related technical activities annually. Their primary mission is the National Defense Program of the Department of Energy: the design, development, and technical support for production, safekeeping, and ultimate destruc-

tion of nuclear weapons. Even with dramatic reductions in the U.S. nuclear weapons inventory under START treaties, there is still an important job for these three laboratories. Dismantling weapons, maintaining those that remain, assuring their security, retarding weapons proliferation, and cleaning up environmental problems at U.S. nuclear sites require the best technical effort DOE can muster. It is unclear, however, how much of the laboratories' current capability these tasks will require. Estimates of these defense needs range from one-third to one-half of the current $3 billion annual budget of the three laboratories.[11] A substantial reduction in laboratory budgets could be a serious blow to the communities around Albuquerque, New Mexico. Sandia and Los Alamos National Laboratories and their associated economic activity comprise the single largest source of income in the state. Anticipating a substantial reduction in support for the nuclear weapons development for which the laboratories were created, politicians are asking, "How can the vitality of these laboratories be sustained and their capabilities put to the best use?"

One possible answer lies in searching for large science projects, such as the project to map and sequence the human genome, which was initiated at Los Alamos, the DOE battery project, to be expanded into the Clean Car project, and in other civilian missions such as advanced transportation, for example the magnetically levitated train, or MAGLEV. Another answer lies in the acceleration of cooperative projects with industry to commercialize technologies in which the laboratories have experience, and in some cases patents.[12] In the summer of 1992, the Senate passed the "Department of Energy Laboratory Partnership Act of 1992," which would have authorized the expenditure of appropriated funds for use in technology transfer projects, had the House agreed.[13] The Clinton-Gore technology policy statement proposed that ten to twenty percent of DOE laboratory budgets be devoted to cooperative R&D with commercial firms.[14]

Senator Bennett Johnston, chair of the Committee on Energy and Natural Resources, introduced the "Department of Energy National Competitiveness Technology Partnership Act of 1993" (S. 473) to address this issue. His bill would confirm the ten-percent goal for all of the Energy Department's multi-purpose laboratories, and would authorize the use of appropriated funds in industrial partnerships. The Senate bill sets out to reduce the time required for negotiating CRADA agreements, and broadens the missions of the laboratories to encompass most of the technical areas in which the laboratories have demonstrated competence. In short, the bill commits the Energy Department, through its laboratories, to a major role in the federal effort to make American industry more internationally competitive. In so doing it may be raising expectations well above what can reasonably be expected from efforts at technology transfer from federal laboratories.

Policy Issues

Authorizing public funding of commercial partnerships with national laboratories in the expectation that this will be effective in supporting competitiveness of American firms raises six questions that are addressed in this chapter:

•The new emphasis on federal support for the technological dimensions of private sector economic performance has resulted in calls for new activities in the area of civilian technology and information infrastructure development, and has cast doubt on the value—or the accessibility to industry—of much of the R&D created by institutions pursuing Cold War missions. How useful are the excess capabilities of national laboratories in support of this new federal role?

•Does the use of public funds in the performance of CRADAs constitute an unwarranted subsidy of selected firms? If so, are there policies that avoid the appearance of market distortion that would still support effective industry partnerships?

•Can the laboratories achieve the goal of having ten percent of their activity in CRADA partnerships without so fragmenting their technical activities that core competencies suffer? If so, does the Clinton-Gore strategy begin the transformation of the Department of Energy into an agency with a very broad technology charter, in effect, a "Department of Science and Technology"? And if not, can the laboratories find a small number of well-defined missions that justify avoiding drastic reductions in staff?

•Can the laboratories manage the new civilian missions well without increased delegation of authority from the agency headquarters? What restructuring of this complex of national laboratories might enhance their effectiveness in the new policy environment?

•Is a system of national laboratories in support of economic progress appropriate? If not, should some of the funds be shifted to universities, or to industry in the form of tax credits for R&D and other indirect incentives?

•What output measures of laboratory success will be used to evaluate their contribution to the economy? If the need is for very diverse research to create useful knowledge, techniques, tools, and standards for use in industry, the utility of this work may be hard to evaluate, especially in the near term. Is the number of CRADA agreements or the amount of counterpart private funds they attract a measure of merit for each of the laboratories?

The Evolution of Technology Transfer Policy

Americans have often criticized the structure of other nations' science and technology enterprises as overly centralized. The French CNRS (National Centers for Scientific Research) and the Australian Commonwealth Scientific Research Organization (CSIRO) are two examples of systems of national laboratories established to support non-military purposes. Both nations have taken major steps in recent years to try to integrate these laboratories more closely into the educational and industrial structure of the country. Americans have traditionally eschewed this pattern for research of economic value, preferring to focus on the universities, which transfer their technical knowledge to industry through training, consulting, and research collaboration. But the U.S. government has embraced the national laboratory system for managing weapons development and acquisition, support for manned space missions, and other federal missions in which the government is the customer for the technology. The national laboratories were primarily meant to support national security and to provide unique, shared facilities for other scientists, most of whom would be in universities.

Spinoff from Government Technology

As discussed in *Beyond Spinoff*, government policy during the Cold War was based on the assumption that spinoff from defense, space, and atomic energy research provided sufficient technological stimulus to keep America's commercial high-tech industry strong. The tacit assumption was that spinoff from work that was justified by federal missions was automatic and cost-free.[15] If so, the economy would benefit from federal expenditures on R&D without the government having to cross the line to industrial policy or find itself "picking winners" from among the available firms to benefit. The government could be blind to the processes through which this government-generated knowledge found its way into the sales catalogs of private firms.

The 1980 Stevenson-Wydler Innovation Act[16] and the federal Technology Transfer Act of 1986, which amended it, recognized implicitly that spinoff is not automatic. Congress charged the agencies to work at technology transfer by forming partnerships with private firms documented by CRADAs. With the Clinton technology policy and with S. 473, the second shoe has dropped; government now realizes that not only is spinoff not automatic, it is not cost-free either. If enacted, S. 473 would authorize the laboratories to spend any or all of their non-defense budgets on R&D to make CRADAs successful.

Obstacles to Effective Technology Transfer in DOE Laboratories

Recent experience has not been encouraging about the likelihood of deriving substantial economic benefits from the transfer of government-owned technologies.[17] Although the National Institutes of Health have moved swiftly to create a substantial number of CRADAs, particularly with rapidly growing, science-based bio-technology firms, the Department of Energy laboratories have faced four major inhibitions:

First, the Technology Transfer Act encouraged CRADAs but assumed that the agency's core mission gives rise to technology useful in meeting both DOE and commercial goals, and that it can be transferred "as is." Second, many firms are highly skeptical that the laboratories possess either the culture (sensitivity to cost and market requirements) or the experience (in process technology, design for manufacturability, and the like) necessary to make a contribution to commercial success. The DOE weapons laboratories grew up in a military arsenal culture, in which the complete life cycle of the products they design and build lies within their own control.

A third obstacle is that laboratory managers are subject to diverse and bureaucratic accountability; their programs are determined by multiple functional offices in DOE headquarters. This makes responsive decentralized decisions (so essential to commercial technology) difficult, and exacerbates the difficulty of negotiating CRADAs in a timely manner. Administrative complexities and reluctance to delegate decision authority to the laboratory management have made CRADA negotiation more tedious and time consuming that most firms will tolerate.

Finally, independent sources such as the White House Science Council have observed that the DOE laboratories are excessively micromanaged from DOE headquarters.[18] Experienced R&D firms identified long-term relationships of mutual trust as an essential element of a successful client-firm relationship. Fear of government inflexibility and intrusion are high among the concerns of firms contemplating CRADA relationships.[19] The effects of Congressional micromanagement are also evident at NIH. The NIH intramural research program (about 20 percent of its $6 billion research budget) is not subject to such severe budget pressure as the defense-related laboratories now anticipate, and collaborations between NIH scientists and bio-tech firms have been quite successful. NIH partnerships serve to support basic science goals by making biological materials available to NIH researchers, and new scientific ideas available to biotechnology firms. The very success of NIH CRADAs has led some members of Congress to urge that NIH use its influence on CRADA partners to persuade them to limit prices charged for drugs developed through partnership with government.[20] While this may be appropriate, the government's attempt to achieve social goals as a *quid pro quo* for CRADAs is almost certain to make them more difficult to negotiate.

When government spinoffs were considered automatic and free, no such demands could be made of firms exploiting government research, once it was placed in the public domain. Once the government becomes a partner in product development, all manner of demands might be made relating not only to prices, but to other elements of social behavior. While firms remain free to "vote with their feet" by walking away from CRADAs with unreasonable demands in them, there is serious danger that Congress may believe its own rhetoric about the commercial value of government science and unintentionally damage the delicate relationship a laboratory needs to be able to negotiate with its industrial partner.

Senator Pete Domenici, co-sponsor of S. 473 with Senator Johnston, recognized the legitimacy of these concerns: "The Department of Energy Laboratories are a huge treasure and storehouse of knowledge and science. My dream is that they could be one of the lead institutions adding to America's ability to apply technology to the marketplace. But their record of traceable new products spun off is so small that one would almost think they are not charged with doing it."[21]

The difficulty of "spinning off" products from technical activities not undertaken for that purpose should not be surprising. Experienced industrial managers testify to the difficulty and risk of commercializing new scientific ideas from their own central R&D laboratories, when the laboratories are working hard at precisely this objective. Even within a single company, such as IBM or GE, moving new technology from corporate research to any of the firm's product development and production divisions requires carefully arranged incentive structures and the full-time attention of very experienced managers. Even then success is very chancy.

This observation does not imply that the laboratories are not supporting their parent agencies with distinction. Many—perhaps most—of them are engaged in supporting the setting of military requirements, developing new technologies to be procured for government use, and assisting procurement offices by monitoring and evaluating contractors. In the case of the nuclear weapons laboratories, they design, manufacture, install, maintain, and eventually dismantle their own products. There is very little need for the complex of intimate relationships with commercial firms that a mission of stimulating commercial innovation would entail.

The Slippery Slope to Mission Fragmentation

Will the pressure to support CRADA partners result in a fragmentation and dilution of the competencies necessary to support core missions? Dr. Robert M. White, president of the National Academy of Engineering, testifying on Senate bill S. 4, noted with reference to both ARPA and the DOE laboratories, "We seem to be heading in the direction of clouding agency roles rather than clarifying them. I'm not sure this will serve the purposes of coherence."[22]

The extraordinarily complex technical challenges of their weapons work, accompanied by generous budgets, have permitted the laboratories to acquire talent and experience in many important technologies, such as high-performance computing, exotic materials, seismic technology and geophysics, and sophisticated design engineering tools. Their mission in energy-related research has acquainted them with a range of transportation and power technologies. Each of these areas of technology touches on additional areas of industrial activity beyond those that have been a direct focus of federal agency interest. A number of them are also the purview of other departments of the government (for example, the Departments of Transportation, Space, and Interior). If the Congress, through legislation such as S. 473, is seen to press the laboratories to broaden their scope of work continuously, the laboratories may lose the technical focus that made them uniquely valuable in the first place.

The current sea-change in federal policy toward the national laboratories was anticipated ten years ago.[23] But concerns about the future of federally funded R&D laboratories have a substantially longer history. This history goes back at least to the "Bell Report," a study commissioned by President Kennedy and conducted for David Bell, director of the Office of Management and Budget, and aimed at establishing criteria for contracting out government work to national laboratories. The most recent study at the level of the Executive Office of the President was carried out in 1983 by the Federal Laboratory Review Panel, chaired by David Packard, for the White House Science Council.[24] Appendix C of the study lists 39 prior studies. In his transmittal letter to the chairman of the White House Science Council, David Packard summarized the panel's concerns:

The Panel's most important recommendations concern the missions and management of the laboratories. First, the parent agencies of the Federal laboratories must review and redefine the missions of these laboratories. At most multi-program laboratories, the research activities could be reduced in breadth, and reconcentrated on those areas most relevant to the missions and of demonstrated excellence. The size of a laboratory must be determined by its mission requirements and by the quality of the work.

Since this conclusion is shared in many previous studies,[25] one appreciates that the current questions about the changing priorities in the federal R&D agenda are being applied to a set of institutions, of which many have not been well focused on important missions even when priorities were not changing as rapidly as they are now. As Representative George Brown said when introducing the Department of Energy Laboratory Technology Act (DELTA) of 1993 (H.R. 1432) on March 23, 1993, "Technology transfer generally needs to be grounded in a mandated technology development effort aimed at satisfying a public mission." This bill takes a somewhat more cautious and disciplined approach than S. 473.

What Assets of DOE Labs Are Relevant to Economic Needs?

The national laboratories, and the Department of Energy laboratories in particular, possess some very important assets that once dispersed would take many years and billions of dollars to reassemble:

•The laboratories are staffed with smart people, more highly educated on average than those in industry. Laboratories like Sandia have a disproportionate number of doctorate engineers, physicists and chemists. Their government tasks have kept most of them at the "cutting edge" of the technologies they practice (although some of these technologies are rather specific to government missions).

•They work in excellent laboratories, better equipped than universities and most corporations. As noted above, many of these government facilities are shared with industry and academic users.

•These laboratories have developed a technical culture that bridges problem solving, technology development and basic research. Thus they can and do move new science into technology with great facility. Unlike most universities, their engineers and scientists are accustomed to working together, creating technologies, not just research papers.

•The labs practice "technology fusion," by building new technologies from a variety of disciplines, a function that only the best companies (and few universities) are organized to perform.[26]

•They possess considerable analytical and theoretical capability, and are generally well equipped to engage in computer simulation and modelling.

•They are accustomed to work on long-term goals with higher technical risks than are common in all but very large companies with assured markets (such as the predivestiture AT&T) .

•In many cases (especially where the work is not militarily sensitive) the laboratories not only have good links to university science, but provide a critical service to academic science by providing access to their unique instrumentation.

•The laboratories are highly motivated to exploit their "new mission in U.S. economic competitiveness" successfully.[27]

Policy Options for New Missions or Downsizing

Whatever changes Congress makes should be made with great care. Great laboratories are more—much more—than their budgets and buildings. The ability to accomplish great tasks depends on an ephemeral quality that might be called the laboratory's culture, spirit, or soul. In a 1983 Argonne Laboratory address I said:

Laboratories do have souls that are expressed in their intellectual traditions, their style, their ambiance, their approach to life. I imagine the souls of human beings are intact from some indefinable early moment in life, but they do not become fully expressed until maturity. So it is with laboratories. The soul of a laboratory is a fragile thing. Long after it is withered due to either abuse or neglect, an institution—like the frog legs in the biology laboratory—continues to twitch its customary twitchings. Outsiders often can't tell the difference for a long time, but people in the laboratory can feel it in their bones.[28]

There is little doubt that America's research engine has a lot of horsepower. The problem, in part, is that the wagon of industrial performance is not well coupled to the public sector part of the research engine; the engine, in turn, is spinning its wheels. How can more than $20 billion in national and federal laboratory R&D and another $20 billion in university R&D be coupled more effectively to the sources of employment and rising quality of life in America?

Given the size of the labs and the scale of their resources, it is very unlikely that any one policy solution to either the search for new missions or preparing for down-sizing will prove adequate or appropriate. Thus the following alternatives are likely to be used in combination.

Continue Current Policies

Make the Stevenson-Wydler approach work in the context of the labs as they are. Continuing on the current path requires no new legislation and leaves lab management free to identify their "dual use" technology and use CRADAs as a mechanism to move technology to private firms.[29] It also imposes no government expectations on the firms looking for cooperation opportunities. This most probable course of action in the near term may, however, also be the least desirable policy. The concept of the CRADA is a sound one if and only if the skills and activities of the national laboratory have a close match to those in the cooperating firms. This seems to be frequently the case in NIH, where the biotechnology industry is still in its science-driven early stage of maturity. How frequently it will be the case in the government's physical science and engineering laboratories is less clear.

Triage

Shrink the labs dramatically and transfer the resources to more effective institutions, using a political mechanism akin to the Defense Department's base closing commission. If one concludes that the public interest would be better served by a shift of resources from national laboratories to universities or to industry (or more likely to some mix of these), the national laboratory budgets are an attractive target. Defense industry executives have been expected to downsize their technical activities; they wonder why the very large national laboratories should be treated differently.

Expand Work Scope

Find new federal missions for the labs within the DOE core mission to nurture new energy technologies and promote more efficient use of energy or find more environmentally benign, energy expending technologies. If Congress interprets the DOE technical mission with sufficient breadth, no new legislation would be needed to authorize the laboratories once again to address such applications. Industrial consortia might be found to locate their joint activities at a national laboratory so that the transfers of technology back to the firms takes place within the staff of each firm. In short, the labs are designed for and experienced at working for a single customer with very deep pockets, a well-defined and stable requirement, and the ability to make all the decisions required to implement the technology developed in the labs. Under the right circumstances this customer could be a consortium of firms, perhaps with other government agencies participating.

Privatize

When the French Thomson company bought RCA, the RCA corporate laboratories were sold to a not-for-profit research organization headquartered in California, SRI International. Thomson then made a contract with SRI to buy back a substantial fraction of the laboratory's effort to support Thomson's development of high-tech television sets for the U.S. market. The rest of the laboratory offers its services for hire on the open market. One could imagine the DOE laboratories receiving authorization to perform "R&D for hire" and thus reducing their dependence on federal support as industry funding rises. Alternatively they could try to sell portions of the labs, with their people and facilities, to industry consortia. The unit costs of work at national laboratories are very high, however, and when added to the transaction costs to firms for buying research from government laboratories the economics of this alternative are unlikely to be attractive.

Reassign Parts of the Laboratories to Other Government Agencies

Parts of national laboratories might be transferred, with budgets intact, to another institutional setting where conditions for effectiveness might be better. One candidate is the Department of Transportation, which has no systems engineering capability on the scale of DOE; another is NIST, which does have a clear mission to address competitiveness issues. This approach would eliminate an intermediate layer of bureaucracy at DOE and would address the problem of a much more rapid buildup of technical resources behind civilian technology programs in Transportation or Commerce than would be possible any other way. The political feasibility of this idea is highly questionable, however, for it requires Congressional committees' willingness to transfer both money and "turf" from one committee to another, which is highly unlikely.

Legislate a New Mission in Support of Industrial Competitiveness

Creation of a new competitiveness mission for the laboratories may be the effect, if not the intent, of S. 473, discussed above. The laboratories would be chartered to respond to industry initiatives and to pursue development of industrially relevant technology, taking advantage of the laboratories' core competencies. This broader mission might look a lot like the broad charter to invest in commercial businesses suggested as the role for the Civilian Technology Corporation proposed by the Research Council's panel chaired by Harold Brown (see Chapter 3).[30]

Effective partnership with industry requires an industry voice in what work is done and how it is carried out. If the projects are developed from the initiatives of consortia of companies (such as the DOE battery project), there may be a useful role for the laboratories and little political objection. If the research is restricted to basic or pre-competitive research, or even to infrastructural or generic research, the benefits will flow indirectly to many users, but the laboratories may have a hard time justifying their huge budgets. If, on the other hand, the labs and their partners select projects with immediate commercial potential, then either competitors may object to what they see as an anti-competitive subsidy and market distortion, or economists and other policy makers may object to replacing the private invest-ments, for which firms should have adequate incentive, with public funds.

Policy Constraints and Mission Opportunities

In government programs such as the Commerce ATP program, the government restricts the use of its funds to public-good technology: in that case defined as pre-competitive, generic technology. If the national laboratories do not have such a constraint, and may use their federal funds to share with partner firms the development of commercial products, the laboratory and its sponsoring agency may find themselves the center of a storm of criticism. This problem can be reduced or eliminated by insisting on full reim-bursement through recaptured earnings, by offering the laboratory's R&D "for hire," or, perhaps, by exacting a *quid pro quo* in the form of some public benefit. But for reasons given above none of these alternatives are attractive.

Alternatively, policy guidelines can limit the perception that public funds are simply being diverted for private gain. These guidelines should not only reduce complaints of unfair, govern-ment-sponsored competition, but also reduce the level of political "earmarking." If CRADAs fail to produce any big winners, there will be few complaints. But given a few successes, serious political problems may arise when particular companies or individuals have reaped large financial rewards, unless it can be established that large spillover benefits or "positive externalities" have also resulted at the same time.

What policy guidelines could assure an appropriate role for the federally funded R&D? As discussed in Chapter 3, economists recognize that where there are demonstrable market failures it may be appropriate for government to compensate for underinvestment by private firms. A common kind of failure is a lack of appropriability of returns to an investing firm for technology which, when diffused throughout industry, provides social returns substantially in excess of cost. Examples of those market failure situations are given in detail in Chapter 3. Those most frequently encountered are, in addition to technology for use by federal agencies in fulfillment of their missions (defense, space, public health, etc.):

•Basic research
•Pathbreaking technologies
•Infrastructural technologies
•High barriers to entry
•Strategic technologies
•Technologies for mixed public/private markets.

These options constitute valid missions for federal agencies and offer quite sufficient scope for the national laboratories. The Los Alamos National Laboratory defines the types of partnerships it undertakes in terms of the first three guidelines above.[31] The work of government biomedical scientists on pathbreaking technologies, for example, has created much of the basis for the biology-based new industry. According to a General Accounting Office (GAO) study, almost 95 percent of all royalty income ($33.9 million out of $35.8 million) received by government from the inventions of its scientists "was earned by employee inventions at five agencies - the Agricultural Research Service, the Alcohol, Drug Abuse and Mental Health Administration, the Centers for Disease Control, the Food and Drug Administration, and the NIH."[32]

Inappropriate Measures of Success

There is a serious danger that the ability to count CRADAs, patents, licenses, and other input data that may be interpreted as giving evidence of successful commercialization of laboratory technology. A Georgia Tech survey of 55 research-intensive U.S. companies found that firms value "access to unique technical resources, knowledge and expertise" more than "the immediate prospect of profit."[33] The 62 industrial division directors who had at least "moderate" levels of interaction with the laboratories had entered into 112 formal, collaborative R&D activities with one or more federal laboratories; 35 of these were CRADAs. The firms ranked technology licensing lowest among sources of value from interactions with the labs. Professor Roessner's conclusion: "There is a danger that too much emphasis will be placed on evidence of tangible economic payoffs (CRADAs, licenses) as measures of success, with insufficient recognition of the value to companies of access to state-of-the-art knowledge and equipment."[34]

The Congress and the administration should back off from setting quantitative goals for the number and size of partnerships with commercial firms, such as the ten to twenty percent of laboratory R&D proposed as the goal in the Clinton-Gore technology policy. All of the laboratories have major activities that are inappropriate for financial matching by private firms acting through CRADAs. The government requires that firms participating in CRADAs provide at least equal funding to that committed by the government laboratory. Thus if the weapons laboratories are devoting half of their activities to the National Defense Program and to basic research, the Clinton policy goal of ten to twenty percent devoted to CRADAs means twenty to forty percent of the remaining projects must be matched by industry. With some $6 billion in the budgets of the DOE general purpose laboratories, this could result in a search for $600 million to $1.2 billion of private industrial R&D funds, or about one percent of all private industrial R&D in the U.S. The bureaucratic interpretation of that goal is too likely to lead to measuring laboratory performance by counting CRADAs, to the detriment of the laboratories' coherence and competence and with little return to the economy. Instead a percentage of the budget, perhaps as much as ten percent, should be identified as an allowable expenditure for the laboratory's participation in CRADAs under laboratory director discretion. Preference should be given to CRADAs in which the initiative comes from industry. Experience with "pushing technology" from research laboratories to product development and production units shows how very difficult this is, even when the lab is part of the firm and a single technical executive is in charge of both units. Some very valuable partnerships may be formed when industry initiates the relationship. The laboratory managers must be given greater authority to negotiate CRADAs quickly, with a minimum of bureaucratic red tape and delay. The provisions on liability and intellectual property proposed in S. 473 could be very helpful in this respect. Laboratory directors should have the authority to approve of CRADAs involving less than half a million of federal funding. The Department should create a "one size fits all" CRADA document, with pre-approved terms and conditions, so that all that is required for a contract is agreement between the parties on the scope of work and the time frame and resources to be committed.

The program as outlined in S 473 may serve to keep the people in the laboratories employed, but, as Congressman Brown notes, it may not be a stable mission, as discussed above. Today the laboratories are operated by contractors, but the contractor does not determine the mission of the laboratory, or control its size and scope of work; this is the responsibility of DOE headquarters staff. A declining level of nuclear weapons R&D, the national budget squeeze, and the difficulty of managing a laboratory struggling to support a wide variety of commercial firms with market interference may all combine to put pressure on the size of the laboratories. They can best adjust to these pressures if the laboratory managements have the authority to restructure the laboratory, diversify its clients, and even privatize or spin off parts of the laboratory to private ownership.

Institutional Innovations

The Congress and the Department should seriously consider incorporating each laboratory as a not-for-profit corporation with its own Board of Trustees, within which is embedded one or more FFRDCs for the DOE's work and that of other federal departments. The physical facilities could be leased to the new corporation for a nominal fee when used for government work; their managements should be allowed to capture and deploy both a management fee and Independent Research and Development (IR&D) and General & Administrative (G&A) funds. This would eliminate an unnecessary level of contract management and collocate responsibility and accountability.

In the longer term, the government will have to think hard about the structure and missions of not only the Department of Energy laboratories, but of NASA as well. Since the energy crisis of the 1970s the Department of Energy has searched for its proper role in its non-defense programs, but its culture bears still the imprint of the Atomic Energy Commission from which it evolved. In the 1970s the new missions in the quest for energy independence taught the Department some hard lessons about how new technologies that depend on commercialization for their value to society should—or should not—be jump-started by government agencies. Solar energy was an example of a path-breaking technology that was pushed into applications long before the technology base made it sufficiently economic. Oil shale development through partnerships showed how vulnerable to shifts in world prices a technology with huge entry costs can be. The Department of Energy must think through the institutional structures and policies through which the mission to make the United States more energy efficient with less environmental burden can be conducted. Ultimately a more decentralized R&D structure than now exhibited in the national laboratories will be needed, one in which industry initiative is given substantially more encouragement. A radical proposal for merging DOE, EPA, NOAA, and the Geological Survey is advanced in Chapter 9.

The National Institutes of Health, in contrast with the DOE weapons laboratories, appear to have been relatively successful with their joint projects with industry, having created 114 CRADAs by 1990, from a start of only ten in 1986. The immaturity of the biotechnology industry, which still draws its sustenance largely from very new science, may be in large measure responsible. A second reason may be that the NIH focus on basic science gave it natural channels for technology diffusion to clinical medicine and applied biology established within the professional research community. The Energy Department laboratories, on the other hand, grew up in a monopsonistic market for its technology: the nuclear weapons and nuclear power programs managed by the federal government itself. Two developments may be emerging that would make the NIH environment look more like the DOE laboratory environment today: the growth of a robust industry dependent on NIH technical support, and new pressures on NIH to take on significant tasks in the national strategy to reduce health care costs.

Similarly, at NIST, where technical interchange with commercial industry has been very close for many decades, the client set is quite diffuse and the channels of information flow are largely created by the professional applied research community: journals, workshop

Figure 4-2 Two Models of National Laboratory Mission and Structure

	Defense, Energy, NASA model	NIH, NIST model
Project structure	Large, interdisciplinary teams	Decentralized, diverse research program
Criteria for federal investment	Government-defined missions: e.g., environment, defense, transportation	Basic, infrastructural pathbreaking science
Mode of industry relationship	Large projects with industrial consortia	Many individual collaborations
Means for technology diffusion and transfer	Uses CRADAs and industrial consortia for technology transfer	Uses professional channels of technology diffusion

conferences, standards organizations, and guest workers from industry at the NIST laboratories. To retain this delicate balance between laboratory research at NIST and outreach to industry, it is important that the NIST laboratory not become administratively subordinate to the new and very rapidly growing new extramural missions given to NIST. ATP contracts with industry, manufacturing technology centers, and industrial extension services with the states should be assigned to a new agency co-located with NIST.

This line of argument suggests two possible models (see Figure 4-2) for a stable relationship between national laboratories and industry (whether in Energy, Commerce, or Health and Human Services): one diverse and decentralized, close to basic science, and using the traditional channels of technical interchange with industry, the other with a few clear, integrated missions conducted in direct collaboration with industrial consortia and justified by unambiguous Congressional mandate to pursue public goals (such as environment, health, safety, or transportation). Either of these models may be stable and effective, but scrambling their attributes could produce political, economic, or technical pathologies. However, common to both models is the necessity of a sufficiently clear and uncontroversial justification for expenditure of federal funds to help private firms make profits, so that the long-term relationships that are essential for CRADA success can be built and left undisturbed by the political turbulence that will always kick up the dust of the industrial policy debate.

Notes

1. The Federal Technology Transfer Act of 1986, Public Law 99-502.

2. William J. Clinton and Albert Gore Jr., *Technology for America's Economic Growth: A New Direction to Build Economic Strength* (Washington, D.C.: The White House, February 22, 1993). See also Clinton-Gore Campaign, *Technology: The Engine of Economic Growth: A National Technology Policy for America* (Little Rock, Arkansas: September 21, 1992).

3. The author is grateful to David Guston for pointing out the significance of these events to both NIH and DOE laboratories; the argument set out here is his.

4. Robert K. Carr, "Doing Technology Transfer in Federal Laboratories," *The Journal of Technology Transfer,* Spring–Summer 1992, pp. 8–28.

5. Government-owned, government-operated laboratories are sometimes known by the acronym GOGO, in contrast with government-owned, contractor-operated (GOCO) laboratories.

6. The definition of an FFRDC given by the NSF in its Science Resources Series is: an R&D performing organization substantially (more than 70 percent) or exclusively financed from federal sources, administered by a university, a private corporation, or a non-profit institution, enjoying a long-term relationship with its sponsoring agency (5 years or more), with its facilities federally owned or funded under federal contract, and its annual budget at least $500,000 per annum. The law governing FFRDCs is Part 35.017, of the Federal Acquisition Regulations, made effective 7 March 1990. Eight federal agencies sponsor 40 FFRDCs of three types: R&D laboratories, study and analysis centers, and systems engineering and integration support. See National Science Foundation, *Federal Funds for Research and Development: Fiscal Years 1990, 1991, and 1992.* Volume XL, NSF 92-322 (Washington, D.C.: National Science Foundation) 1992, p. 10.

7. An anomaly in the government's categorization is that the Applied Physics Laboratory, a large applied military research laboratory, is operated by Johns Hopkins University and is considered a part of the university for government accounting purposes.

8. The DOE laboratories are of two types: the weapons laboratories, including Livermore Research Laboratory, Los Alamos National Laboratory, and the Sandia Laboratory, and the multifunctional research laboratories, such as Fermilab, the Berkeley Radiation Lab, Argonne National Laboratory, Brookhaven National Laboratory, and Oak Ridge National Laboratory.

9. *Facing Toward Governments: Nongovernmental Organizations and Scientific and Technical Advice* (New York: Carnegie Commission on Science, Technology and Government, January, 1993).

10. The growth of NIH followed the pattern set by the Public Health Service by the addition of non-uniformed government scientists and physicians. However, eighty percent of NIH research funds are spent outside the huge laboratory complex in Bethesda, Maryland.

11. John Nuckolls, director of Livermore, stated that nuclear weapons research, as a percentage of Livermore Lab R&D, fell from 48% in 1988 to 36% in 1991; *Science,* November 22, 1991. The same article reported that Siegfried Hecker, director of the Los Alamos National Laboratory, said that weapons research employment has dropped by one-third in six years.

12. Incentives were established to encourage the formation of CRADAs in the 1986 Stevenson-Wydler amendments, called the Federal Technology Transfer Act of 1986 (Public Law 99-502).

13. S.B. 2566.

14. Clinton and Gore, *Technology for America's Economic Growth.*

15. John A. Alic, Lewis M. Branscomb, Harvey Brooks, Ashton B. Carter, and Gerald L. Epstein, *Beyond Spinoff: Military and Commercial Technologies in a Changing World* (Boston: Harvard Business School Press, 1992), p. 9.

16. Public Law 96-480, October 26, 1980.

17. Alic, et al., *Beyond Spinoff,* chap. 3.

18. David Packard, panel chair, *Report of the White House Science Council: Federal Laboratory Review Panel* (Washington, D.C.: Office of Science and Technology Policy, Executive Office of the President, May 1983), p. 9. George E. Brown, Jr., Chairman of the Science, Space, and Technology Committee of the House of Representatives, noted in September 24, 1992, hearings, that the very diffuseness of mission of the laboratories encourages micromanagement.

19. Roger W. Werne, *U.S. Economic Competitiveness: A New Mission for the DOE Defense Programs' Laboratories,* UCRL-ID-112185 Rev. 1 (Livermore, Calif.: Lawrence Livermore Laboratory, November 1992), quoting Battelle, SRI International, Southwest Research Institute, and others he surveyed.

20. The Congress has applied pressure to the National Cancer Institute to extract commitments from CRADA partners such as Bristol-Myers-Squibb, developer of the anti-cancer drug Taxol, to limit its prices for the drug. The CRADA agreement in this case declares the government's "concern that there be a reasonable relationship between the pricing of Taxol, the public investment in Taxol research and development, and the health and safety needs of the public."

21. *Science,* May 13, 1988, p. 874.

22. Robert M. White, "The Emerging National Technology Strategy," Testimony to the Senate Commerce Committee hearing on S.4, March 25, 1993.

23. "A case can be made that a restructuring of the Federal research and development strategy, and its institutional underpinnings, is overdue. The driving force for this case is that in many people's minds, the major challenge facing American science and technology is economic rather than deriving from narrower Federal government missions such as space, defense, and energy." Lewis M. Branscomb, "National Laboratories in the Nineties," Director's Special Colloquium, October 12, 1983, published as an offprint by Argonne National Laboratories.

24. Packard, *Report of the White House Science Council: Federal Laboratory Review Panel.*

25. In the late 1960s the President's Science Advisory Committee (PSAC) asked Dr. Albert Hill of MIT to lead a panel to study the national laboratories and recommend ways to keep their missions up to date and improve their performance. He came back to the next meeting with a report consisting of a single transparency, on which appeared a matrix displaying a dozen prior studies of national laboratories and corresponding recommendations. The matrix was a forest of "x's," since every study had come to the same set of conclusions. His recommendation was, "Implement the recommendations of any of the prior studies." Most of the Packard report recommendations are repetitions of these earlier analyses.

26. Fumio Kodama, "Technology Fusion and The New R&D," *Harvard Business Review,* July–August 1992, pp. 70–78.

27. Werne, *U.S. Economic Competitiveness.*

28. Lewis M. Branscomb, address at Argonne National Laboratory, 1983, pp. 14–15.

29. Dr. William Happer, Director of the Office of Energy Research and Science Advisor to the Secretary of Energy, notes that in 1992 DOE doubled the number of CRADAs from 83 to over 200, representing $300 million in cooperative R&D, 60 percent of which is paid for by private industry. William Happer, testimony to the Committee on Science, Space and Technology, U.S. House of Representatives, September 24, 1992.

30. This interpretation has been contested by Dr. William Happer, DOE Director of Energy Research in the Bush Administration, both of the sponsoring Senators of S-473, and the directors of DOE laboratories.They note that the laboratories developed technical expertise in areas such as high performance computing, global climate change, nuclear medicine, advanced materials and manufacturing, transportation and space technologies. However, the pursuit of commercialization of any of these technologies in partnership with commercial manufacturers, and the laboratories' pursuits in support of a DOE mission in nuclear weapons development, are two quite different missions. If S. 473 authorizes public investment in any part of the technology required for commercial success, the goals of those projects will necessarily constitute a broadening of the DOE mission. See transcript, March 18, 1993, hearing before the Senate Committee on Energy and Natural Resources.

31. Kay V. Adams, presentation at the Fiftieth Anniversary symposium of the Los Alamos National Laboratory, April 16, 1993.

32. General Accounting Office, *Technology Transfer: Barriers limit royalty sharing's effectiveness* (Washington, D.C.: United States Government Printing Office [U.S. GPO], December 1992), pp. 33–35.

33. David Roessner, "Patterns of Industry Interaction with Federal Laboratories," to be published in *Research-Technology Management* (published by the Industrial Research Institute) and in the *Journal of Technology Transfer* in 1993. The work was carried out in cooperation with the Center for Innovation Management Studies, Lehigh University, the Industrial Research Institute, and the Department of Energy.

34. This conclusion is also confirmed in the survey of University-Industry Research Centers; see Wesley Cohen, Richard Florida, and W. Richard Goe, *University-Industry Research Centers in the United States,* Center for Economic Development, Carnegie-Mellon University, Report to the Ford Foundation, June 1992, to be published in book form in 1993.

Appendix 4-A Federal Obligations for FFRDC R&D and R&D plant by Agency, FY 1992 ($ millions)

Sponsoring Federal Agency	Total	Industrial	Universities	Other
Dept. of Energy	$4,249	$1,775	$2,252	$222
Dept. of Defense	1,707	485	799	423
NASA	886	—	884	2
NRC	152	87	34	31
NSF	123	0.6	122	0.1
HHS (mostly NIH)	73	19	33	21
Dept. of Transportation	14	—	—	14
Dept. of Labor (mostly Bureau of Labor Statistics)	7	7	—	—
Dept. of Interior (Bureau of Mines)	3	3	—	—
Total All Agencies	$7,217	$2,377	$4,126	$714

Figures are rounded and may not total exactly. FFRDCs sponsored by Commerce, Justice, State, Treasury, Archives, and ACDA are less than $1 million each. Source: National Science Foundation, *Federal Funds for Research and Development: Fiscal Years 1990, 1991, and 1992*, Volume XL, NSF 92-322 (Washington, D.C.: National Science Foundation, 1992), p. 55.

Appendix 4-B   FFRDCs with R&D obligations greater than $100 million, FY 1990 ($ millions)

FFRDCs	Total	DoD	DOE	HHS	NASA	NSF	Other
Industry contract							
Bettis Atomic Power	332	94	237				
Idaho Nat'l Engin.	188		155				32
Knolls Atomic Power	291	31	259				
Oak Ridge Nat'l Lab	435	172	236	2			25
Sandia Nat'l Lab	718	118	573				27
Savannah River Lab	153	153					
University contract							
Argonne Nat'l Lab	252	11	233	1		0.5	6
Brookhaven Nat'l Lab	213	9	177	6		0.5	21
Lawrence Berkeley Lab	240	3	224	12			
Lawrence Livermore Lab	654	168	480	2	1		3
Fermilab	158		158				
Jet Propulsion Lab	699	82			616		
Lincoln Labs	210	210					
Los Alamos Sci. Lab	629	138	483	5			3
Stanford Linear Acc.	115		114	1			
Not-for-profit contract							
Aerospace Corp.	116	113			2		
Mitre Corp.[1]	164	164					
Pacific NW Lab[2]	120	2	92	11			16
Total all FFRDCs[3]	6,336	1,493	3,895	63	622	107	155

1. Budget shown is the Air Force C³I FFRDC.
2. Pacific Northwest Laboratory is operated by Battelle Memorial Institute.
3. These totals are for all FFRDCs, including those smaller than $100 million a year, such as the CTI.
Source: National Science Foundation, *Federal Funds for Research and Development: Fiscal Years 1990, 1991, and 1992,* Volume XL, NSF 92-322 (Washington, D.C.: National Science Foundation, 1992), p. 55. Numbers are rounded and may not add. Contractor categories identify the institution holding the contract for the laboratory's operations: industry, university (or consortium of universities), and not-for-profit institution, respectively.

Appendix 4-C   Federal Obligations for Intramural R&D by Agency for FY 1992 (only agencies with greater than $100 million shown; estimated in rounded $ millions)

Agricultural Research Service	575
Forest Service	147
National Institute of Standards and Technology (NIST)	144
National Oceanic and Atmospheric Administration (NOAA)	272
Dept. of the Army	2,169
Dept. of the Navy	2,805
Dept. of the Air Force	2,930
Defense R&D Agencies (DARPA)	1,798
T&E: Deputy Undersecretary of Defense	186
Department of Energy (DOE)	449
Alcohol, Drug Abuse, Mental Health	267
Food and Drug Administration (FDA)	116
National Institutes of Health (NIH)	1,485
Geological Survey	303
Federal Aviation Agency (FAA)	184
Veterans Administration	219
Environmental Protection Administration EPA).	104
National Aeronautics and Space Administration (NASA).	2,613
Smithsonian Institution	112
Total	17,645

For source of data see Appendix 4-B.

5

Information Technology and Information Infrastructure

Brian Kahin

This chapter looks at the special characteristics of information technology and the emerging concept of information infrastructure, which plays a central role in the Clinton/Gore technology policy. We look at the Internet as the driving paradigm of information infrastructure and how the history of the NSFNET network exemplifies the problem of infrastructure privatization and commercialization. Finally, we review the components of the administration's information infrastructure vision, which articulates a three-tier investment strategy and draws telecommunications policy and information policy under the purview of technology policy.

The Nature of Information Technology

Information technology is a uniquely enabling technology which supports the development and diffusion of other technologies. It is extending with increasing coherence and logic into a seamless information infrastructure that embraces the fabric of social and economic enterprise, rationalizing and quickening institutions and industries. As information technology draws hitherto distinct industries—publishing, entertainment, telecommunications, broadcasting and cable, computers—into a common environment, information infrastructure has become a rallying point for strategic public investment. However, despite economic arguments for public investment, it is not easy to design strategies for public investment in this fast-moving, competitive, convergent environment.

Information technology is often viewed as a very rapidly evolving technology like biotechnology and other fast-moving technologies. But information technology is ultimately distinguished by its versatility and mutability. Just as the computer is a "universal machine" that can be applied in infinite ways to an infinite variety of processes and systems, both physical and abstract, information technology is the quintessential enabling technology. A software application such as a spreadsheet can be a tool for research, product development, manufacturing, and commerce. It can provide a platform for the development of additional tools. It can provide a reference framework or "infrastructure" for a particular enterprise.

Standards play an especially important role in information technology because of the complexity of information processes and the need for system-wide interoperation, two characteristics which would otherwise work against each other. Complexity is easy to achieve because the costs of complexity are in the design rather than the manufacture. Once the process is designed, a product can be stamped out as an integrated circuit, or, in the case of software, reproduced for the negligible cost of the electronic media. Hence information technology, software in particular, has been able to emulate and automate physical and mental processes at a very reasonable cost.

The very richness of this functionality and the many possible ways of encoding it make it difficult to connect and interoperate different processes, however internally coherent they may be. And so there is a strong need (if there is to be a competitive market) to establish standardized interfaces between processes to enable assembly, integration, and extension of systems. As they gain in scale, coherence, acceptance, and market demand, systems become platforms—stable reference points on which further applications, systems, even new platforms can be built. Platforms can be more or less proprietary, such as the Intel 80x86 series of processors, the IBM-standard PC, and Microsoft's MS-DOS.[1] Or they can be public domain, as is the case for most consensus standards. The Internet— where interoperability is the *sine qua non*—is the epitome of a public technology platform. As platforms grow, and especially as they accommodate and interoperate with other platforms, they begin to look like infrastructure: a pervasive, organizing whole, supporting a wide range of applications for a wide range of users. This general-purpose infrastructure draws established institutions, markets, laws, and practices into contact with new resources, users, and markets with scale economies and standards. Participation in the common infrastructure gives established institutions tools and a framework to redesign their internal infrastructure.[2]

Federal investment in basic research in areas such as supercomputing, massively parallel computing, advanced algorithms, and high-speed networking appears similar to investment in other pre-competitive technologies. But the very breadth of the impact, especially in the case of general-purpose networking and computing, sets information technology apart from the technology policy mainstream. The pervasiveness and speed of technological change and the new economics of digital information are profoundly challenging to a wide range of institutions. Information technology is sometimes superficially perceived to be a solution to deep-seated institutional problems—or, alternatively, a red herring pushed by computer zealots. In reality, it is a demanding tool that may introduce new challenges or reveal other problems; it can embed obsolete and inefficient processes, but it can also compel rationalization, rethinking, and redesign.

Information technology also affects a broad range of institutions and personal activities quite apart from the technology-intensive parts of the economy. For most people, technology means information technology, because information technology has for many years been the most visible, palpable agent for change in work, education, and entertainment. Public familiarity with the impact of television, cable, fax, and personal computers feeds the vision of universal switched (in contrast to broadcast) broadband services that would transform the way that every home and business is connected to the world.

Accordingly, information technology touches many policy areas traditionally quite distinct from technology policy: privacy and security, intellectual property, freedom of speech, management of government information, as well as telecommunications policy—

even such stable areas of the law as contract and theft. It can be especially disruptive in areas where courts have to weigh competing rights, values, and interests. In such areas, the lack of experience and legal precedent may breed uncertainty, anxiety, overreaction, and skewed public perceptions.

Defining Information Infrastructure

While information technology is commonly understood to mean computers, software, peripherals, and networks, the concept of information infrastructure is difficult to circumscribe. The term has been used only since the explosive growth of microcomputers and computer networking in the early 1980s, and many would argue that information infrastructure is unique to digital information—that it makes little sense to talk of an infrastructure for analog information. Analog information is passively carried by a telecommunications infrastructure; digital information, because it can be logically linked and perfectly replicated and because (in the form of a computer program) it can organize and manipulate other digital information, creates its own infrastructure.

Interoperating data networks are plainly at the heart of information infrastructure, but does it extend to the computers on the networks? Are information technology standards part of the infrastructure? What about organizations and individuals using the network, the digital information resources available over the networks, and the users of the digital information who may not be network users themselves? We do not think of cars, and certainly not their drivers, as part of the transportation network, but computer networks, computers, software, data, and even users interoperate at a high level in complex ways. They may even substitute for each other and shape each other, and this rich systemic interdependence looks very much like infrastructure.

There have been many attempts to define information infrastructure, but most are representative of three distinct perspectives, singly or in combination. From one perspective, information infrastructure is an aggregation of systems for carrying information, much like transportation infrastructure provides for the carriage of people and other physical objects. From another perspective, information infrastructure is the aggregation and organization of digital information in a form intelligible and useful to human beings. From a third perspective, information infrastructure is the logical fabric inherent in any system or enterprise that is monitored or controlled by information technology; by extension, it is also a larger fabric, based on common standards and networks, that enables interoperation and exchange of data among different systems or enterprises.[3]

These three perspectives embody different visions of information infrastructure, which may be described individually as *telecommunications infrastructure, knowledge infrastructure,* and *integration infrastructure.*[4] They reflect historically distinct communities of interest, driven by different principles and market forces and addressed by different policy traditions. But they now share an interest in the widespread diffusion of digital information technology, systems, and services in a standards-based environment—i.e., a common infrastructure. As we shall see, the Clinton-Gore administration appears to appreciate the interrelationships between these different aspects of information infrastructure and is addressing them in concert through its technology policy.

Telecommunications Infrastructure

Telecommunications infrastructure, which has long been recognized and analyzed as infrastructure, embraces the physical networks and switches that underlie all communications beyond the home or office. In most countries, telecommunications infrastructure has been provided by agencies of the government; in the United States, it has been provided by the private sector but regulated at both state and federal levels. As with the analog telephone system, the vision of the future telecommunications infrastructure is driven by mass-market economics and the ideal of universal service.

The prospect of universal switched broadband is enticing, but the principle of universal service is clearly difficult to apply in the present data environment where there is continuous, rapid technological change with many levels of capability and many kinds of functionality available at any given time. Furthermore, universal service was mandated by regulatory policy in a single-provider environment. Regulators are now preoccupied with how to provide a level playing field for a growing variety of players, including cable companies, competitive access providers (CAPs), and personal communications services (PCS), as well as local telephone companies.[5] In addition, the stakeholders themselves have different technological options, which can be discouraged or encouraged by regulatory policies. For example, there has been debate over whether local telephone companies should be encouraged to deploy narrowband ISDN (Integrated Services Digital Network), a relatively inexpensive upgrade from conventional telephone service. ISDN could accelerate the development of some applications, but it might also erode demand for broadband services.[6]

Knowledge Infrastructure

Knowledge infrastructure embraces the organization and processing of information for access, interpretation, and use. Its principal communities of interest include research, higher education, information-intensive professions, and electronic publishing. The High Performance Computing and Communications Program, especially the National Research and Education Network component (NREN), has become the principal locus of federal investment in knowledge infrastructure. In addition, government information, especially government-generated scientific and technical information, contributes substantially to the resource base of knowledge infrastructure.[7]

Early instantiations of knowledge infrastructure include the machine-readable bibliographic databases produced by the American Chemical Society (Chem Abstracts) and others (often with help from the National Science Foundation); the MEDLARS databases developed by the National Library of Medicine; the national bibliographic utilities, OCLC and RLIN;[8] the massive electronic law libraries, Lexis and Westlaw; and online vendors, such as Dialog and BRS, that have developed software and systems to distribute the electronic databases of others.

The vision of knowledge infrastructure that evolved with the rise of distributed computing in the 1980s is that of an open environment which provides a marketplace for accessing information in an infinite variety of forms from an infinite variety of sources with search processes chosen by the user. The Internet offers a platform for this marketplace by allowing for multifunctional interconnection among distributed sources worldwide. This platform has recently been enhanced by the development and deployment of higher-level protocols for distributed information retrieval.[9]

Policy debate in the development of knowledge infrastructure often centers on the interaction of public investment and private investment. Public investment is common in early stages when a market is not yet established and transaction costs are high. But public investment usually carries with it an expectation that the service will become self-sufficient, either supported by users on a cooperative cost-recovery basis or, if the market is large enough, provided competitively by the private sector.[10] But this transition is seldom easy or obvious, as the debate over the commercialization and privatization of the Internet has shown (see below).

Integration Infrastructure

Integration infrastructure is a common environment of information standards and interoperating networks that enables the automation and integration of processes, systems, and whole enterprises within and across institutional boundaries. Integration infrastructure is critical to visions and goals of agile manufacturing, enterprise integration, concurrent engineering, advanced manufacturing, product data exchange, etc., all of which are driven primarily by the competitive concerns of the manufacturing sector. However, integration infrastructure is not limited to actual manufacturing processes; it encompasses the entire product life cycle, which includes R&D, product design, maintenance and support, as well as contracting and marketing processes.

The principal federal program for integration infrastructure has been CALS, the DoD's Computer-aided Acquisition and Logistics Support initiative. CALS encompasses a broad set of standards-development activities undertaken in conjunction with NIST and the private sector. CALS seeks to develop dual-use standards that will enable DoD to build on the civilian technology base while implementing a DoD-wide platform for automating weapons design, procurement, deployment, and maintenance. Thus CALS supports integration between the defense economy and the civilian economy, between DoD and its contractors (and subcontractors), and among the fragmented and bureaucratized procurement and logistics offices within the different services of the United States military. In the cutting-edge area of standards for the exchange of product data, CALS is joined by the Commerce Department's National Initiative for Product Data Exchange, which is designed to build broad industry involvement in standards development and implementation.

CALS, by virtue of its breadth, its direct investment in the standards development, and the lure of procurement billions, has become a focal point for industrial standards development in the United States and abroad. Indeed, CALS has provided a pragmatic approach to a national standards policy that has often been viewed as nonexistent.[11] However, CALS remains under-recognized and suboptimized because it is viewed as a specific DoD program and because it lacks a network orientation and an aggressive deployment program. For example, in 1991 the public database of CALS materials was moved from NIST, where it was Internet-accessible, to the National Technical Information Service, which, two years later, is still not on the Internet.

The handful of CALS Shared Resources Centers are presently islands which, like the Manufacturing Technology Centers funded by NIST, need to be integrated into a systematic plan for industrial extension. The new Technology Reinvestment Project (headed by ARPA but involving four other agencies) has components for dual-use and defense conversion extension services that may help provide the necessary context. In addition, the new Federal Coordinating Council for Science, Engineering,and Technology (FCCSET) interagency initiative in the FY94 budget, the Advanced Manufacturing Technology (AMT) program, provides among other things an investment framework for integration infrastructure, including standards, integration models, assembly networks, and deployment. Some of this funding will likely support MCC's Enterprise Integration Network (EINet), which provides a Internet-based platform for industrial networking, including deployment of standards and other common resources.

The Internet as Paradigm for Information Infrastructure

In many respects, the Internet provides the central organizing paradigm for information infrastructure. Its roots are in the research community and it has been driven by the knowledge infrastructure vision of access to distributed information resources. But it has moved this vision out of the world of research and higher education toward the goal of universal connectivity that characterizes telecommunications infrastructure. Once thought of as merely a prototype for the broadband networks of the future, it has clearly become a major evolutionary pathway by virtue of its vast and growing user base and the extensive applications it supports.

While the links to integration infrastructure are less apparent, Internet standards development stands as a remarkable and instructive case of standards development and deployment by users. The network itself serves to engage a large number of extremely talented individuals without the expense and tedium usually associated with standards committees. Electronic Mail (e-mail) allows rapid comment and reiteration with an instantly retrievable record. Drafts and official standards alike are instantly available without cost. As a result, Internet standards have been thoroughly tested by a wide variety of users, and the Internet's TCP/IP (Transport Control Protocol/Internet Protocol) suite has been extensively implemented in private networks and commercial software. Despite longstanding assumptions (held by NIST and other government agencies) that TCP/IP was at best a temporary solution, it has proved eminently workable and has spawned many higher-level applications. By contrast, the OSI (Open Systems Interconnection) protocol suite favored by the traditional telecommunications standards bodies (and a product of anticipatory committee-based standards development) remains incomplete, untested, and lacking a critical mass of applications.

The Internet is defined functionally rather than institutionally. It is the set of interconnected networks that support the interoperation of three basic functions: remote log-in, electronic mail, and file transfer. It is not limited to TCP/IP networks; networks supporting OSI or other protocols are part of the Internet if they interoperate with the predominant TCP/IP Internet through protocol conversion.

In the United States alone, forty to fifty Internet networks, apart from the federal agency networks, offer Internet services. What cooperation exists is a result of a multitude of *ad hoc* arrangements, some involving federal agencies, especially NSF, some entirely private. This lack of central organization is sometimes perceived as a triumph of unregulated pluralism and sometimes as an invitation to disaster. For example, the rapid growth of commercial addresses is creating concern that solutions are needed to resolve address space shortages and routing problems, but while solutions have been suggested, it is not clear who can or should take responsibility for implementing them. The only institution roughly congruent in scope with the Internet as a whole is the Internet Society, an international professional society which now houses the standards development bodies, the Internet Engineering Task Force and the overseeing Internet Architecture Board. But the Internet Society has no say over the operations of any individual network.

The Internet remains bedeviled by misconceptions. It is still often perceived simply as an electronic mail network which public funding makes available free to universities. In fact, the Internet is high-speed, multi-functional infrastructure leveraged in a number of ways:

•Enormous private investments in computers and private networks, especially local area networks. These investments have leveraged inter-network connectivity and contributed to the rapid commercialization of TCP/IP internetworking technologies;

•Carrier investments in fiber optic networks driven primarily by the market for conventional telephone service and leased lines for private networks. The Internet consists of logical overlay networks, riding on leased lines and requiring little capital investment other than switches.

•Federal and private investment in very high-speed networking technologies. This includes both testbeds and the NSFNET backbone which has been available as a free resource to network service providers (but not to end-user networks).

•Statistical multiplexing which is characteristic of the packet-switched technology. Although congestion can occasionally slow performance on the Internet, the overall costs are much lower than for circuit-switched communications.

These factors, together with low marginal costs for capacity, result in inexpensive fixed-fee capacity-based pricing at the institutional level. At a large university, where there may be five to ten thousand Internet users, annual connection to the Internet through a mid-level network costs on the order of five to ten dollars per user per year. The cost is usually simply absorbed into institutional overhead. If the NSF network subsidies were removed entirely, the average cost to institutions would rise no more than 30 percent.

The Internet, NREN, and Information Infrastructure

Another common misperception is that the Internet is a single network. The Internet is a *metanetwork*—an international set of autonomous networks which support a common level of functional interoperation. Similarly, the National Research and Education Network is not a new government network. It is a federal program, part research and part enhancement of existing federal Internet investments.

The High Performance Computing and Communications Program, including the NREN, began as a FCCSET initiative under the Bush administration prior to the passage of the High Performance Computing Act,[12] but under Bush the program remained focused on basic research on computing and networking technologies and on meeting the networking and supercomputer needs of scientific researchers. This focus on basic research did not change with the passage of the High Performance Computing Act of 1991, despite the Act's broad language portraying the NREN as a path to a universal broadband infrastructure. Although the Act required the NREN to be developed and commercialized in collaboration with the telecommunications, computer, and information industries, the Bush administration did little to implement such collaboration. The Act mandated a private sector advisory committee for the HPCC program as a whole, but President Bush had not made a single appointment to the committee by the time he left office a year later.

Even within this narrow focus, there has been continual confusion over the distinction between research, research infrastructure, and the seeding of broader infrastructure. This confusion is an product of an extremely dynamic and volatile technological environment and the need for a compact, salable vision. The image that then-Senator Gore introduced and which remains in the Clinton-Gore technology policy is the "information superhighway," the NREN as an analog to the Interstate Highway System. However, this image is confusing because it suggests that fiber will be laid as a public works project to create a single government-funded network which would not be available until the fiber is in place.[13]

The infrastructure investment component of the NREN, as distinct from the networking research component, is essentially a continuation and enhancement of the current NSFNET program in coordination with other agency networks. (The overall implementation plan is described as the Interagency Interim NREN, or IINREN.) NSFNET, too, suffers from misconceptions. It is not, as the name suggests, a network owned and operated by NSF. It is a NSF program which funds:

•a highspeed backbone service (the "NSFNET backbone"), which is provided free of charge to connecting networks for traffic that meets NSF's acceptable use policy (AUP). That is, the traffic must be in support of research or education;

•certain mid-level networks—autonomous regional, state, and supercomputer-centered networks, most of which function as co-operatives. NSF subsidizes a shrinking number of these networks to provide regional connectivity and connect to the NSFNET backbone;

•initial connections to the mid-level networks by colleges and universities—in the form of one-time grants of no more than $25,000.

Although it is common to speak of NSFNET as encompassing the backbone service and the connecting mid-level networks, this is misleading in several respects. The backbone is used by many networks (including commercial providers) that have never received NSF funding, and by mid-level networks that were initially subsidized by NSF but no longer receive direct support. Even the networks that NSF subsidizes are free to determine their own policies for internal traffic (or traffic with other consenting net-

Figure 5-1   Expenditures for NSFNET Program: 1988–91 ($ millions)

Fiscal Year	1988	1989	1990	1991
Backbone	2.7	3.6	4.9	8.8
Regional Grants	3.6	7.1	7.1	7.3
Connections	0.2	0.5	0.7	1.4
Other	1.2	2.3	2.4	3.5
Total	7.8	13.6	15.0	21.0

Source: National Science Foundation, Division of Networking and Communications Research and Infrastructure

works), although all traffic that makes use of the NSFNET backbone must be in support of research or education.

The NSFNET backbone was originally a dedicated network provided under a cooperative agreement between NSF and Merit, the Michigan Higher Education network. In September 1990, a nonprofit corporation, Advanced Network and Services (ANS), was formed by IBM, MCI, and the State of Michigan Strategic Fund to operate a private backbone out of which the same level of backbone services would be provided to NSF. Under this arrangement, ANS remained free to sell capacity beyond the NSF requirements. This privatization of the backbone through the modification of a cooperative agreement created considerable controversy, especially after ANS spun off a for-profit subsidiary in May 1991 and sought agreements with regional mid-level networks to handle commercial traffic. Other commercial providers accused ANS of taking unfair advantage of its central position in the NSF-subsidized infrastructure, and hearings were held before the House Subcommittee on Science in March 1992. The Subcommittee Chairman, Congressman Rick Boucher, requested an investigation by the NSF Inspector General, which was released to the public along with the response from the NSF Director in April 1993. Although critical of NSF's procedures, the Inspector General's report concluded that the provision of commercial services from the same facilities that provided NSFNET backbone services was acceptable because ANS was to return a portion of the commercial revenue to a cooperatively administered fund for research and education infrastructure, and because NSF had later announced that the ANS backbone could be used by other commercial providers on the same basis.

This controversy over the commercialization of the Internet was, from another perspective, a measure of NSF's success in seeding infrastructure. Statistics for traffic on the NSFNET backbone from 1987 on show continual dramatic growth at rates of eight to fifteen percent per month. NSF's liberal policies resulted in a decentralized Internet that rapidly turned competitive in areas of high demand, such as Silicon Valley. The open, networked standards-development process on the Internet, which NSF and the other federal agencies supported, spurred the growth of the global Internet and TCP/IP technologies. The interoperability and broad interconnectivity of the Internet in turn stimulated the development of higher-level protocols for handling distributed information resources and numerous applications and services. By early 1993, there were an estimated 4–7 million Internet users in the United States alone, and the overwhelming majority of new site registrations were for commercial hosts.

Along with this rapid growth came a growing constituency for advancing and broadening the Internet to serve the needs of libraries, K-12 education, health care, manufacturing, and so on. But the Bush administration continued to view federal programs in high-performance computing and the NREN as two things: First, basic research; second, as a tool or utility for the "Grand Challenges" on the nation's science agenda, such as global climate change and mapping the human genome.[14] This view served the interests of the agencies involved in the HPCC, because it supported their established programs and constituencies and did not impose new burdens and demands that they were not equipped to handle.[15]

Meanwhile NSF extended the cooperative agreement with Merit that was to have expired in November 1992, as NSF sought to design a new infrastructure model that accommodated other agency interests and the growth of the commercial Internet. A solicitation concept was put out for comment in May 1992 and by the close of the comment period in August, NSF had received input from 44 interested parties, including a number of regional Bell operating companies (RBOCs) and other telecommunications providers. The widespread interest in the solicitation demonstrated conventional carriers' concern that it would set a lasting precedent and possibly provide a critical mass of applications around which the switched broadband networks of the future would take shape.

Private investment in Internet had raised questions about how best to promote efficient services, competition, and further private investment. Under NSF's original three-tier model, funding policies supported cooperative "mid-level" networks, some of which were partially funded by state governments. It seemed to suggest a territorial division of the nation into benevolent monopolies, but that was not to be. The first commercial providers, AlterNet and PSINet, appeared in late 1989, targeting areas of high demand, and there was occasional competition among the nonprofit regional providers, who had come to understand that any formal turf division would violate antitrust law. An ensuing controversy between ANS and its commercial competitors, represented by the Commercial Internet Exchange (CIX), made it clear that there could be no preordained monopoly at the top level of the hierarchy. The Internet, especially the commercial Internet, was becoming a mesh of networks with no sustainable distinction between long distance (backbone) and regional providers.

NSF found itself caught between two views of the backbone. The CIX, which included commercial providers but not ANS, argued that the requirement for backbone services was a requirement for a production network, which could be satisfied using proven, "off-the-shelf" technology. Others felt that the backbone should be cutting-edge technology at the heart of a "research network" which embodied a pre-commercial technology.

NSF responded by unbundling the concept behind the old cooperative agreement, so that backbone services would be separated from a routing authority. A new concept, Network Access Points, was introduced. NAPs would provide a neutral ground for interconnecting commercial and noncommercial networks. While the highspeed backbone would still be restricted to research and education uses, the NAP would be "AUP-free" and, except maybe in the beginning, would not be subsidized. At one point, NSF appeared to accept the production network argument by announcing that it would fund at least two backbone service providers.[16] But by early 1993, NSF had moved away from an infrastructure model back to a research-only model of a high speed backbone network that would connect the NSF-funded supercomputer centers for parallel processing and other experimental uses. Not only would the highspeed backbone be limited by the AUP, but routine uses such as e-mail and file transfers would be phased out.

While the NSF has offered assurances that it will help the regional networks make the transition into this arrangement, it will no longer wholly underwrite interregional connectivity for research and education by providing a free backbone. However, NSF will still be providing varying direct assistance to certain regionals that are competing increasingly with unsubsidized services. The only legitimate reason for continued subsidies at the network level would appear to be exceptional circumstances such as high line-mileage to user ratios in certain geographic areas.[17]

There has long been pressure on NSF to move the subsidy down to end-user institutions or even grant recipients, so that network service providers would chosen by the users rather than by NSF. On the one hand, this would help remove disincentives for private investment and stimulate a competitive market for network services. On the other hand, having the backbone available as a free resource for all network service providers minimized transaction costs and helped rationalize and focus demand around a common set of protocols and management tools. Once the infrastructure has jelled, however, the arguments for efficient, user-driven markets grow stronger.

But NSF has been justifiably leery of putting its investment entirely in the hands of end-user institutions. This would raise difficult questions about qualifying expenditures (or qualifying networks) and eligible institutions. There would either be an enormous administrative overhead in adjudicating among hundreds, eventually thousands, of proposals, or the NSFNET would have to take on the characteristics of an entitlement program, something NSF had no experience in operating and which would be certain to raise political eyebrows. The alternative of tacking on funds for network use to individual NSF grants would raise complex accounting issues (since networking expenses are generally treated as fixed overhead) and would fail as an infrastructure program by rewarding only the elite few who are already receiving NSF support.

Perspectives of the New Administration

The approach of the Clinton-Gore administration—clearly foreshadowed by the vice president's agenda as a senator—has been to subsume NREN into a broad vision information infrastructure to be overseen at a high level in OSTP and the National Economic Council. This has been a near-term necessity because of delays inherent in the nomination and confirmation process and the reassessing of ongoing programs. However, it will be necessary over the long term to ensure inter-agency coordination and to maintain a coherent vision and perspective over a difficult and unwieldy set of issues.

The Office of Science and Technology Policy of the Clinton-Gore administration sees five strategic areas for advancing information infrastructure:

•the inter-agency High Performance Computing and Communications Program (HPCC) as authorized under the High Performance Computing Act (HPCA) of 1991;

•the "Grand Applications" (digital libraries, health care, K-12 networking, and advanced manufacturing) addressed by Gore as a senator in S. 2937, "The Information Infrastructure and Technology Act of 1992";

•demonstration projects to advance Internet connectivity for schools, government agencies, and nonprofit entities at the local level;

•telecommunications policy; and

•information policy, including government information, intellectual property, privacy, and security issues.[18]

The first three are areas of strategic investment. The first reiterates the administration's commitment to the HPCC Program.[19] The second investment area embraces the "Grand Applications" articulated in late 1991 by the Computer Systems Policy Project, in

S. 2937, by the Information Infrastructure and Technology Act of 1992 ("Gore II") introduced by then-Senator Gore the following July, and by two bills in the 103rd Congress, S. 4 and H.R. 1757, introduced by Senator Ernest Hollings (D-S.C.) and Congressman Rick Boucher (D-Va.). The Clinton-Gore program, articulated in *Technology for America's Economic Growth*, promises to: "create an information infrastructure technology program to assist industry in the development of the hardware and software needed to fully apply advanced computing and networking technology in manufacturing, in health care, in life-long learning, and in libraries." The administration views the HPCA as providing adequate authorization for the program (at least as cover for any supplemental funding); however, it has also worked closely with Boucher in designing H.R. 1757. K-12 education, which was identified as a Grand Application in "Gore II," is treated separately in the section on initiatives on education.[20]

The third investment item is clearly infrastructure development:

Provide funding for networking pilot projects through the National Telecommunications and Information Administration (NTIA) of the Department of Commerce. NTIA will provide matching grants to states, school districts, libraries, and other non-profit entities so that they can purchase the computers and networking connections needed for distance learning and for hooking into computer networks like the Internet. These pilot projects will demonstrate the benefits of networking to the educational and library communities.

The administration requested $64 million in supplemental funding for this program ("on-ramps" for information highways), which makes it clear that this is not aimed at only one-of-a-kind demonstration projects. Presumably it would be similar to NSF's existing Connections program, which has provided one-time grants of up to $25,000 to bring educational institutions onto the Internet. Funding for other entities would have to come under the Public Telecommunications Facilities Program (PTFP) administered by NTIA. Although PTFP primarily funds public broadcasting facilities, the authorizing language is broad enough to include Internet access.

Under the administration's plan, information infrastructure development would also be driven by increasing dissemination of government information. This program is in some respects a cross-over between investment policy and regulatory policy. It is discussed below under regulatory policy.

The two remaining components of the Clinton-Gore technology, telecommunications policy and information policy, are major policy areas in which the government plays a very different role than in the strategic investment components. As stated in *Technology for America's Economic Growth*:

Government telecommunication and information policy has not kept pace with new developments in telecommunications and computer technology. As a result, government regulations have tended to inhibit competition and delay deployment of new technology. For instance, without a consistent, stable regulatory environment, the private sector will hesitate to make the investments necessary to build the highspeed national telecommunications network that this country needs to compete in the 21st Century. To address this problem and others, we will create a high-level inter-agency task force within the National Economic Council which will work with Congress and the private sector to find consensus on and implement policy changes needed to accelerate deployment of a national information infrastructure.

The proposed Task Force would include the Departments of Commerce, Justice, and State, but not the Federal Communications Commission, which, because of its special status as an independent agency, cannot be directly involved in executive-branch policy development. This is nevertheless a step toward integrated policy development in an area that has been plagued by jurisdictional conflicts, and it is the first time that concerted telecommunications policy-making has been elevated to the White House level since the White House Office of Telecommunications Policy was divested to the Commerce Department (as NTIA) twenty years ago.

The interagency task force under the National Economic Council as part of a larger program for information infrastructure development could provide leverage for transforming telecommunications policy by, in effect, assimilating telecommunications policy to technology policy. This may help break the present political gridlock among industries (cable, telephone, newspaper) that have different regulatory traditions and different competitive characteristics.

The regulatory categories set up by the Communications Act of sixty years ago have become less and less meaningful as the established monopolies and oligopolies have been pressed by new technologies and competition. The FCC, continually buffeted between Congress and the courts, has become a reactive agency in an increasingly deregulated and technologically volatile environment. Its most significant policy development functions center around spectrum allocation. But spectrum allocation is increasingly an international issue (where the State Department plays a leading role) and a technology issue. As a technology issue, it should be addressed as part of a policy development process cognizant of the many technologies that are not subject to traditional telecommunications regulation.

This convergence of technology policy and telecommunications policy has been forced in part by the success of agency policies that have stimulated the development of the Internet as a switched public metanetwork entirely independent of the established regulatory environment. Indeed, the relatively limited reach of government regulation in the United States and, in particular, the highly competitive market for leased lines seem to account for the tremendous growth of Internet connectivity and services in the United States compared to Japan and Europe.[21] However, as the Internet grows, the decisions of NSF and the other agencies are assuming a quasi-regulatory force. The controversy over the privatization and commercialization of NSFNET backbone makes it clear that the architecture of IINREN (Interagency Interim NREN) is likely to strongly influence the emerging broadband infrastructure. It therefore merits some of the economic and policy scrutiny to which traditional telecommunications regulation has been subject.

The fifth element of the Clinton-Gore strategy focuses on information policy and technology. It is even more remarkable that the administration intends to address information policy along with telecommunications policy under the interagency Task Force. Despite the presence of the word "Information" in its name the National Telecommunications and Information Administration has rarely attempted to confront information policy issues. But as noted earlier, OSTP counts information policy as a separate element of the administration's five-part agenda on information infrastructure.

The attempt to address information policy as part of technology policy is most probably the most ambitious part of the administration's plan. It is more significant than integrating tele-communications policy and technology policy, because competi-tive and technological forces are already transforming telecommunications policy (albeit more slowly than might be desirable). Unlike telecommunications policy, which has a reason-ably coherent literature and regulatory history grounded in eco-nomic analysis, "information policy" is a catch-all for a variety of issues that lack a common theoretical foundation.[22] Rhetoric, ideology, and legal argumentation characterize much of the policy discourse.

There are several loose clusters of issues in information policy. One centers on First Amendment, privacy, and security issues; it is connected to issues of national security, export control, and law enforcement. A second area of information policy is management of government information, including access to information un-der the Freedom of Information Act (FOIA), and proactive dis-semination of government information. *Technology for America's Economic Growth* specifically addresses the latter, which can also be seen as a public investment issue;

Promote dissemination of Federal information. Every year, the Federal govern-ment spends billions of dollars collecting and processing information (e.g., the economic data, environmental data, and technical informa-tion). Unfortunately, while much of this information is very valuable, many potential users either do not know that it exists or do not know how to access it. We are committed to using new computer and networking technology to make this information more available to the taxpayers who paid for it.

This does not mark any radical change from existing federal policy. OMB has been engaged for the past two years in a revision of Circular A-130, governing dissemination of federal government-gathered and generated information. Congress has also been wrestling with pricing practices for dissemination of government information; there, proposed language has been close to that of the draft revision of A-130. In both, pricing of government information products would be no more than the marginal cost of dissemina-tion. While not always precisely defined, marginal cost pricing is justified on economic efficiency grounds.

The Copyright Act precludes domestic copyright for works of the United States Government, and, under both proposed legislation and the A-130 draft, agencies would not normally be permitted to assert contractual controls over the redistribution of government information. Without such controls, the market price of government information would normally be driven toward marginal cost whatever the government's price. Since the marginal cost of information in the digital environment is low and rapidly growing lower, this should increase the flow and utilization of information that is not otherwise controlled. The easily availability of such information should, in turn, drive demands for computers, networks, and other elements of information infrastructure.

Tools such as "gopher" and WAIS, now available over the Internet, provide access to the information resources that have been distributed, but other than the major research agencies, few agencies have mounted significant resources on the Internet. In principle, this is becoming easy and inexpensive to do, but it is unclear whether the agencies are capable of doing an adequate job of mounting information (which could be accessed through common tools) or whether instead it is advisable for the National Technical Information Service (NTIS) or the Government Printing Office (U.S. GPO) to set up special systems (e.g., "Fedline" or "WINDO").[23] NTIS, which has been revitalized in the last two years, has better relationships with the agencies and is considerably more agile than GPO; it is, however, charged by statute with fully recovering its costs of operation and has been slow to come to grips with the economics of network distribution. GPO, however elephantine, has close ties to Congress and a strong relationship with the library community through the GPO Depository Library Program.

One area of controversy that has arisen in recent years is whether the prohibition against copyright protection for works of the United States Government should apply to software produced by government employees in national laboratories. Congresswoman Constance Morella (R-Maryland) introduced legislation in the last Congress that would have permitted copyright for such works when created under a cooperative research and development agreement (CRADA) with private industry. However, given that CRADAs can be created at will with little oversight (as long as there is no government funding of the private sector partner), this begs the larger question of what sort entrepreneurial activities the national labs should be free to engage in on their own initiative. In any case, despite the support of technology activists in Congress, the bill was stalled by broad opposition from publishing industries and library organizations.

A third area of information policy issues, in principle complementary to government information, is intellectual property. In many respects, intellectual property in itself is both as complex and as jurisdictionally fragmented as telecommunications regulation. It, too, involves regulation of markets; however, unlike telecommunications regulation, public policy issues are typically framed as legal issues. Despite the impact on competitive markets and constant concern about "strengthening intellectual property," there is little economic analysis in intellectual property debate (other than calculation of damages in individual cases). Furthermore, while the disparate approaches to telecommunications policy by Judge Greene in the AT&T breakup, the FCC, the states, NTIA, the State Department, the Justice Department, and Congress are focused by a common set of statistics and economic principles, the several intellectual property regimes (patent, copyright, trade secret, trademark) have historically been quite independent of each other.

The administration quite properly views intellectual property (at least the federal regimes, copyright and patent) as a technology policy matter central to information infrastructure. Indeed, copyright regulates the flow of privately created information and patent regulates the flow of innovative technology. Intellectual property has been nominally on the agenda of the Technology Administration in the Department of Commerce, but the Patent and Trademark Office is independently situated within the department, and recent administrations have focused almost exclusively on the special problems of protecting American intellectual property abroad. Policy development, such as it was, was centered on trade negotiations, especially multilateral arrangements under the Uruguay Round of the GATT.

The greatest challenge for the administration will be to inject accountability into the patent system, which has evolved with very little oversight in recent years but which should be closely integrated into a coherent technology policy. The globalization of trade and the push for multilateral agreements has made patent law harmonization a major issue, and there has been a slowly growing consensus that the United States should accept the first-to-file system that prevails in the rest of the world as part of a negotiated package to harmonize national patent laws. Because there has been considerable resistance to moving to a first-to-file system, especially from university researchers and small inventors, an Advisory Commission on Patent Law Reform was set up in 1990 to help pave the way for legislative action. However, the area of inquiry that drew the most response from the Commission's Request for Comments[24] was not the first-to-file question but software patents (or, as the U. S. Patent and Trademark Office prefers to call them, "patents on computer-related inventions"). This is a complex and controversial area that offers ample opportunity for debate, but which testifies to poor policy development and bad management on the part of the government, as well as to the capture of the patent system by its high priests, the patent bar.[25]

When Congress instituted the Court of Appeals for the Federal Circuit (CAFC) in 1982 in order to end the inconsistency of patent rulings issuing from the different Appeals Court Circuits (and the forum-shopping that this caused), it did not intend to radically increase the power of patentees at the expense of the free market. But as CAFC decisions were rendered, usually by judges who were members of the patent bar, that is what happened, whether the issue was scope of patentable subject matter or the weight accorded to patent examiner assessments of non-obviousness. In the case of software, where the USPTO (United States Patent and Trademark Office) has lacked a database of prior art and the vast majority of patents are thought to be invalid for lack of novelty,[26] the results have been disturbing. Patentees with broad non-technical patents covering business methods, educational methods, and other abstract processes, and armed with a strong presumption of patent validity, can pursue mere users of software who are easily frightened by the extraordinary costs of patent litigation. The power and reach of patents creates real risks for the development of information infrastructure, because the complex interdependencies of the infrastructure may easily be disrupted by the unexpected appearance of a patent.[27] The potential scope of the problem was unintentionally enlarged by the Process Patent Amendments Act of 1988, enacted at a time when Congress (along with most of the software industry) was unaware that software patents were being granted. Aimed at foreign manufacturers, the Act makes it an infringement to use a product created with a patented process. Hence, merely reading a spreadsheet created with an infringing spreadsheet program would presumably violate the patentee's exclusive right to control the use of a product created by its patented process.

A major problem is that there is no policy development function for the patent system; the Patent and Trademark Office is a virtually autonomous office within the Department of Commerce; it employs no economists; and it interacts almost exclusively with those who benefit directly from increased perception of the importance of patents and the expansion of the patent system into new areas.[28] Patent law has become so much the province of specialists that the Supreme Court has grown reluctant to review patent cases, effectively yielding control to the patent brethren in the CAFC.

Patent law is premised on a *quid pro quo*: the inventor receives a seventeen-year monopoly in exchange for disclosing the invention to the public. In theory, therefore, patent publication should be an important component of a diffusion-oriented technology policy. However, under the Reagan administration, patents were viewed principally as an incentive mechanism and there was little interest in the diffusion and utilization of patent information. While this improved somewhat under Bush, there is continued tension with private providers of patent information who had earlier taken advantage of patent office indifference.

Such problems could be addressed by making the Patent and Trademark Office part of the Technology Administration in the Department of Commerce, making sure that resources are available for informed policy development (including economic analysis), working with the affected industries. The latter is most critical for industries on the fringe of the patent system, such as software development, multimedia publishing, financial services, education, and commercial services. Ideally, policy development for copyright should be incorporated under the Technology Administration as well, although Congress would undoubtedly be reluctant to move the Copyright Office from the Library of Congress.

Conclusion

This chapter can only hint at the complexity of the policy environment that the administration faces on information technology issues. As this is written, the administration has announced an encryption technology developed by the National Security Agency for commercial use—the "Clipper Chip"—only to have the computer industry and others reject it as inadequate.[29] Although the government does not play such an active role in copyright as it does in patent, copyright law suffers from ambiguity and confusion at a number of critical points. With the proliferation of new publishing models, there are new and unanswered questions about how the results of government-funded research should be disseminated. The list goes on.

Nevertheless, this is the first administration to understand the root interconnectedness of information, technology, and infrastructure. It has implicitly scoped out a policy development agenda that may take years to fully develop, but it has made a strong start in bringing the issues together. It has also drawn appropriate links between information infrastructure and other major components of its technology policy: education and training; advanced manufacturing technology; and making government more efficient.

The importance of information technology in making government more efficient is clearly expressed in *Technology for America's Economic Growth*, but it is unclear how a program would be implemented. Should leadership come from the Office of Management and Budget, or is there a need for an aggressive hands-on program built on the principles of enterprise integration that are on the research agenda? The only agency that has the resources, need, and (one hopes) the motivation to lead such a program is the Department of Defense. Indeed, a substantial initiative was already building within the DoD under Paul Strassmann, Director of Defense Information in the Bush administration. Strassmann was able to bring the CALS initiative within the purview of his Corporate Information Management (CIM) initiative and was able to articulate a compelling vision for reengineering DoD management processes through the design and implementation of a DoD-wide information infrastructure.

If anything is missing from the administration's information infrastructure vision, it is an articulation of the role that DoD can and should play in developing integration infrastructure for the federal government. If Clinton hopes to "reinvent government" on a broad scale, he must demonstrate that the DoD, having a conservative culture but also a need to downsize and integrate into the civilian technology base, can be reinvented as an agile national security enterprise tailored to the post-Cold War era. There are difficult questions about how to draw on the private sector expertise that will be necessary to accomplish this, but the issue must be confronted.

Here, as elsewhere, the administration must develop the economic and policy case that builds on the extraordinary range of justifications for public investment in information infrastructure. These range from support for agency missions (especially DoD, NASA, DOE, and NSF) and support for basic research (e.g., gigabit networks) to providing a suitable safety net analogous to the public library and lifeline telephone service. In between lie a number of economic justifications based on the public-good character of information and basic research, network externalities, producer and consumer information deficiencies, economies of scale and scope, asset specificity, and transaction cost barriers, as well as the federal government's ability to leverage investments by other levels of government and by the private sector. These justifications apply most forcefully in the early stages of infrastructure development, but in a dynamic, stratified technological environment, there are always some communities and technologies in the early stages of development.

At the same time, rapidly changing technologies and market conditions militate against government intervention. Policy-oriented agencies generally lack the resources to understand and assimilate new technology in sufficient depth. Even the Patent and Trademark Office has difficulty mustering the resources to address new areas of technology.[30] Agencies which fund advanced research, such as NSF and ARPA, do a better job of keeping on top of the technology but often lack the expertise to appreciate the economic and political complexities of infrastructure development.

Information infrastructure itself is part of the answer. Networks and enterprise integration strategies can help government agencies re-engineer their systems and processes and bring them into closer communication with their private sector constituencies. More generally, a fully developed and deployed information infrastructure can facilitate inter-sector coordination just as it can facilitate inter-enterprise integration. In particular, this coordination should include redefining the concept of technology transfer by using the instantaneous feedback and other market-like functions that a fully realized infrastructure can provide. Indeed, the impact that computer networks have had on the operation of financial markets hints at the opportunities for market-oriented technology licensing. New mechanisms for ensuring anonymity and comfort with the network should be explored, such as the use of trusted intermediaries as brokers. As a facilitator and monitor of markets, the infrastructure should eventually be able to signal not only unmet demand but also market failures that the public sector might address. In short, the infrastructure must advance beyond the exuberant information-sharing of the early Internet to become an adaptable framework for interleaving market mechanisms, joint enterprises, traditionally integrated firms, and, where necessary, public sector investment.

Notes

1. "Proprietary" can mean many things in information technology. It can mean that the product (or interface) is unique to one company. Or it can mean that the product is protected as intellectual property—trade secret, patent, or copyright. In fact, different aspects of a computer program may be protected by different forms of intellectual property. Even if particular information technology is protected as intellectual property, the proprietor may choose to license it, sometimes at nominal or no cost if the proprietor is seeking the advantages of broad market penetration. A proprietor may encourage complementary but not substitute products (e.g, Apple encourages third parties to develop software for the Macintosh, but discourages hardware clones).

2. The DoD, which now looks to CALS and the civilian technology base to help reform its internal processes, is the best example in the federal government of how participation in a common infrastructure enables redesign of institutional arrangements.

3. This ambiguity is reflected in the term, "enterprise integration," which means both intra-enterprise and inter-enterprise integration.

4. Senator Jeff Bingaman, "The National Information Infrastructure," *Roll Call*, March 8, 1993, p. 26. Bingaman's vision is partitioned into the technological

elements—communications, information, and computing—rather than the institutions and demand drivers that characterize the three perspectives.

5. The trend has been to pursue the goal of universal service with more specific, economically rationalized concepts such as targeted "life-line" service.

6. There has been much literature and exhaustive debate on the relative technological, regulatory, and marketing advantages of cable companies and telephone companies for providing broadband services to the home. Telephone companies have been inhibited by regulatory policies that favor low-cost residential telephone service; the provisions of the Modified Final Judgment (ordering the divestiture of AT&T), which included proscriptions against manufacturing and, until recently, against provision of information services; uncertain demand for ISDN as well as more advanced switched digital services; and continued rate-of-return price regulation for basic services coupled with very limited allowances for plant depreciation. With prospects for radical upgrading of the public network uncertain, much of the earnings of the RBOCs (regional Bell operating companies) have been invested in businesses abroad, which have exploited proven skills and technologies while promising relatively high financial returns.

7. Library programs within the Department of Education has been a source of funding for library networking and data-sharing projects, although these have lacked the leverage and impact of NSF's Internet investments.

8. OCLC was originally the Ohio Computer Library Center. It is now universally known as "OCLC," which nominally stands for the Online Computer Library Center. RLIN is the Research Libraries Information Network, operated by the Research Libraries Group.

9. These protocols include WAIS (Wide Area Information Service), gopher, archie, and WWW (World Wide Web).

10. Private firms provide services more efficiently, but may do so with the expectation of a monopoly-like position facing inelastic demand. Because of the unique character of many information resources, such as specialized databases, it may be difficult to challenge an established provider. Indeed, a comprehensive provider may find itself with special obligations under antitrust law as an essential facility. One of the issues in an ongoing lawsuit between Dialog and the American Chemical Society is whether the Chemical Abstracts Service is an essential facility which ACS must make available through Dialog as well as through ACS's own online system, STN.

11. See the OTA report, *Global Standards: Building Blocks to the Future* (Washington, D.C.: Office of Technology Assessment, March 1992). One of the strategies discussed is development of an information infrastructure for standards. See p. 26ff.

12. See the "Blue Book," *Grand Challenges: High Performance Computing and Communications*, which accompanied the President's 1993 Budget.

13. The misconception was also fueled by lobbying efforts such as those of Opt In America, which claimed that government regulation was inhibiting investment in optical fiber. In fact, carriers invested heavily in fiber for trunk lines throughout the 1980s resulting in substantial excess capacity (which has meant low costs for the networks of the Internet). Both voice carriers and cable companies have found that the life cycle costs of fiber were less than conventional wiring and are installing fiber in new construction of shared facilities (i.e, almost everywhere but the "last 100 feet").

14. See the "Blue Book," *Grand Challenges: High Performance Computing and Communications,* and the "Teal Book," *Grand Challenges 1993: High Performance Computing and Communications,* issued as supplements to the president's FY92 and FY93 Budgets.

15. This view is evident in OSTP's report to Congress in December 1992 on a set of NREN-related policy questions posed by the High Performance Computing Act. Although the report goes beyond the questions to provide a status report on the NREN program, there is little in it that reflects the expansive visionary language in the High Performance Computing Act.

16. NSF support for two backbone service providers would have followed the model of the FTS 2000 procurement for federal telecommunications services, which was designed to be awarded to two providers who would have to coordinate with each other.

17. Cooperative regional networks may have a continuing advantage in areas outside the urban areas in which the interexchange carriers have points of presence (POPs). Rather than require newly connecting institutions to lease lines all the way back to the POPs, regional cooperatives can facilitate line-sharing arrangements, even though the cost of the line is initially the responsibility of the newly-connecting institution.

18. To a large extent, these five areas are documented in William Clinton and Albert Gore, Jr., *Technology for America's Economic Growth* (Washington, D.C.: The White House, February 22, 1993). However, the five-part description is based on a presentation by Mike Nelson, OSTP, to the Forum on Information Infrastructure convened by the Harvard Information Infrastructure Project at the University Club in Washington, D.C., on April 7, 1993.

19. *Technology for America's Economic Growth,* p. 17 calls for "Implementation of the High-Performance Computing and Communications Program established by the High-Performance Computing Act of 1991 introduced by Vice President Gore when he served in the Senate. Research and development funded by this program is creating (1) more powerful super computers, (2) faster computer networks and the first national high speed network, and (3) more sophisticated software. This network will be constructed by the private sector but encouraged by federal policy and technology developments." Unfortunately, the statements about the network, while working to allay private sector fears that the Federal Government will be laying fiber, resurrect the impression that there will be a

single highspeed network (the NREN?) and that it will come into being only when "constructed" sometime in the future. ANS could certainly argue that its implementation of the 45 mbps backbone was "the first national high speed network." Presumably, it all depends on what is meant by "high speed."

20. "Access to the Internet and developing NREN will be expanded to connect university campuses, community colleges, and K-12 schools to a high-speed communications network providing a broad range of information resources. Support will be provided for equipment allowing local networks in these learning institutions access to the network along with support for development of high-performance software taking advantage of the emerging hardware capabilities." *Technology for America's Growth*, p. 14.

21. Kenneth Flamm and Frederick Weingarten, presentation on comparative networking initiatives, Forum on Information Infrastructure, April 7, 1993.

22. See, for example, Jane Yurow, ed., *Issues in Information Policy* (Washington, D.C.: NTIA, 1980).

23. Fedline is an operating prototype at NTIS which allows gateway users to dial up a bulletin board to gain access to certain agency databases. WINDO was a proposed Government Printing Office gateway which would have been authorized by H.R. 2772, introduced in the 102nd Congress.

24. 56 *Federal Register* 22702 (May 16, 1991).

25. The report of the Advisory Commission on Patent Law Reform, which was issued in September of 1992, is of dubious value because the 14-member committee lacked the broad representation required by the Federal Advisory Committee Act. The controversy surrounding software patents was glossed over with rhetoric on the historical value of patents, and the Commission avoided questions raised by the impact of the Court of Appeals for the Federal Circuit on the presumption of patent validity and the likelihood of finding infringement.

26. Glen Self, vice president for research at EDS, estimates that 90 percent of software patents are invalid on the basis of novelty alone. Despite the USPTO's inability or unwillingness to evaluate the scope of software patenting, EDS has assembled a massive database of patents which include software claims. See U.S. Congress Office of Technology Assessment, *Finding a Balance: Computer Software, Intellectual Property and the Challenge of Technological Change* (Washington, D.C.: U.S. GPO, 1992), pp. 24-25 and accompanying footnotes.

27. A well-known instance is when the distributors of MIT's X-Window system, which MIT had made freely available to the world as a standard for multiple-session displays, were surprised by AT&T's claims that the system infringed on its "backing store" patent (No. 4,555.775). See John Markoff, "Patent Action On Software By AT&T," *New York Times*, February 26, 1991, p. D1. Although in this case the patent had issued, its possible application to the X-Window system was unknown to MIT and the members of the X Consortium.

28. Significantly, the president has nominated Bruce Lehman, a copyright specialist, as Commissioner of Patents and Trademarks. In general, Commission-

ers of Patents have been patent lawyers, and Lehman's nomination has drawn fire from the patent bar, who view him as lacking the requisite experience.

29. Export controls on encryption technology continue to be a sore point with the computer and software industries because the restricted technologies (especially the RSA public key system) are widely available throughout the rest of the world and foreign competitors have been able to exploit the market freely.

30. The difficulties the USPTO has had with software patents are often attributed to this generic problem. But for thirty years, until the early 1980s, the Patent Office resisted granting patents on software for such practical reasons. Then it changed policies without investing the resources to keep current, let alone the resources needed to develop a database to make up for the missing thirty years.

6

Industrial Extension and Innovation

Gene R. Simons

I believe that we have to do a better job of encouraging . . . all firms to be involved and more competitive in the work place—but I believe that we have to help those that haven't had a chance to really develop.
Representative Xavier Becerra (D-California) on his appointment to the House Committee on Science, Space and Technology.

The amendment I proposed eliminates one of the portions of the competitiveness bill that deals with the funding of more than 150 technology centers. . . . The centers are absurd. . . . You don't need more than 30 of them, maximum.
Representative Martin Hoke (R-Ohio) on his appointment to the House Committee on Science, Space and Technology.

Congressional interest in implementing a federal technology policy during the Bush administration resulted in several small programs under the Department of Commerce. One of these, the Manufacturing Technology Centers (MTC) program, is based on the premise that smaller manufacturers are the foundation of U.S. industry. The designers of the MTC program defined the technological improvement of the smaller manufacturers as a necessary precursor to the resurgence of U.S. manufacturing. Seven MTCs have been funded during the past four years to transfer advanced manufacturing technology to smaller manufacturers. In addition, the federal government has provided limited planning grants to the state-based technology extension programs. The election of President Clinton promised a substantial increase in activity in these areas. This chapter examines the rationale for a federal role in industrial extension and suggests approaches to meet the technology needs of the smaller manufacturer in a cost-effective manner, while linking to existing programs at the state level.

During the presidencies of Reagan and Bush (1980 to 1992), the manufacturing sector of the U.S. economy lost ground to both imports and the recession, resulting in the loss of 1.8 million manufacturing jobs and a negative trade balance which peaked at $167 billion in 1987.[1] President Bush did, however, establish the Technology Administration in the Department of Commerce, which focused on the National Institute of Standards and Technology (NIST) and its role in facilitating technology transfer in industry. This was referred to as "technology policy," rather than "industrial policy," with the administration eschewing programs that provided direct support to individual firms in favor of investment and tax incentives.

Despite political nervousness about government assistance to manufacturing businesses, the industry itself seems quite ready for a bolder approach. The president of the National Association of Manufacturers, Jerry J. Jasinowski, described the association's position:

Federally funded R&D more relevant and useful to meeting the competitive needs of the nation should be considered on its merits, not on the basis of ideology.... Generic manufacturing R&D efforts, focused on base-building technologies and processes rather than specific products should be promoted. . . . The federal government should also assist—not control—state and local governments in their efforts to promote local technology development. . . . Government and industry should expand their support for manufacturing-related research activities as well. In short, the best and latest R&D must be applied to manufacturing to make and keep U.S. industry the most productive, cost-effective, and market-responsive in the world.[2]

As 1992 came to a close, the federal government began to respond to this argument.

The focus of President Clinton's manufacturing initiative, however, is the same as that of President Bush, that is, to provide technology support for the smaller manufacturers that supply 60 percent of the components used in domestic manufacturing. The underlying premise is that the key cause of the decline in U.S.

manufacturing competitiveness has been the inability of smaller manufacturers to implement manufacturing process technologies and techniques which would support high-quality, low-cost production, while allowing for rapid changes in product design.[3] There is widespread acceptance of this belief. The large American original equipment manufacturers (OEMs) have increasingly turned to offshore suppliers because they are often considered superior to U.S. firms in price, quality and service.[4]

The Bush administration's position on technology or industrial policy appeared to shift in 1992, when the undersecretary for technology, Robert White, issued the Technology Administration's "Strategic View."[5] This report proposed expanding the MTC program to 30 large centers and 100 small centers over the next eight years. Bipartisan support developed when Governor Clinton promised in his platform to expand the MTC program to 170 "market-driven" centers and to provide support for improving the state industrial extension operations.[6] In October 1992, Senator Bingaman proposed in the Department of Defense budget revisions to spend $540 million in Fiscal Year 1993 on state and federal initiatives in this area.[7] Based on the bipartisan support implied by these actions, we anticipate that support for smaller manufacturers through industrial extension will be substantially expanded during 1993.

President Clinton's February 1993 statement[8] outlined a broad technology policy which featured the development and dissemination of advanced manufacturing technology to rebuild U.S. manufacturing and thus create needed jobs. The Defense Advanced Research Projects Agency (DARPA) was renamed ARPA and given the responsibility for dual-use technologies, that is, products and services that can be used in both defense and non-defense applications. In *Beyond Spinoff*, the authors argued for a proactive defense policy in the dissemination of defense created technology into the civilian sector.[9] Dropping the "D" from DARPA symbolizes the attempt to initiate this policy shift.

The special attention given to manufacturing extension and its role in improving the productivity of small manufacturers in Clinton's policy statement was reinforced in March 1993, when ARPA issued a broad agency announcement describing the Technology Reinvestment Project (TRP), which was funded with $472 million of reprogrammed FY93 DoD funds.[10] The stated mission of TRP is to "stimulate the transition to a growing, integrated, national industrial capability which provides the most advanced, affordable, military systems and the most competitive commercial products." TRP combines the efforts of the Departments of Defense, Commerce, and Energy with the National Science Foundation and the National Aeronautics and Space Administration. It provides three major thrusts: Technology Development (which is 45 percent of the funds), Technology Deployment (45 percent of funds), and Manufacturing Education and Training (10 percent of funds).

The Technology Deployment section's main component is the Manufacturing Extension Program (budgeted at $87 million), which is aimed at increasing the competitiveness of smaller manufacturers by "stimulating the introduction and use of advanced technologies to improve both products and manufacturing processes." Technology Deployment also includes extension enabling services (linking the extension services), supplier development programs, and technology access services to federal laboratories.

This effort is directly based on the success of the Manufacturing Technology Centers Program, which managed to survive earlier administration opposition and between 1989 and 1992 established seven extension centers under NIST for the purpose of transferring advanced manufacturing technologies to small to medium-sized manufacturers to improve their productivity and competitiveness. With a 1992 budget of only $17 million, the MTC program was deemed to be too small to have any real impact on the industrial sector.[11] During the first three years of operation, only 1200 small manufacturing establishments were assisted.[12] And compared to similar programs in other countries, such as Japan's program funded at $500 million per year, the U.S. federal role was quite small. In addition, there are 42 state-funded industrial extension programs in 28 states spending $83 million per year, which service about 2500 firms per year.[13] Therefore, we can estimate that the combined services of state and federal extension programs in 1992 reached less than 1 percent of the manufacturing establishments in the United States: at a cost of $100 million per year.[14]

Even the definition of "advanced" manufacturing technology has been changed by NIST during the past four years. Originally, the phrase referred to leading-edge state-of-the-art technologies such as robotics and laser technology. After three years of experience with the MTCs, NIST modified this definition to "off-the-shelf best practices."[15] This change was important to the extension agent, who was able to provide limited technical support to the smaller manufacturer, and to the smaller manufacturer, who was dependent on the technology vendor for technical support. This change also reflects a shift from the supply-side policy, where the focus is on the transfer to industrial firms of technology created by federal laboratories, to a demand-side policy aimed at meeting the actual needs of the smaller firms, which are better served by existing, proven technologies along with implementation assistance.

Thus, the first months of 1993 have represented a turning point in the federal role in manufacturing and indicate that a manufacturing policy is indeed a goal of the Clinton administration. This chapter discusses the premises on which the administration's policy is based and the role of advanced manufacturing technology in American competitiveness. We also suggest an implementation strategy which, if the TRP is a real indicator, is already underway.

Public Policy Issues

Chapter 1 discusses the public issues relating to technology policy, highlighting the complexity of the relationship between technology policy and economic policy. Industrial extension presents an additional set of public policy issues that must be addressed by the federal funding agents before a comprehensive national program is put in place.

The first issue is the working relationship that must be developed between the federal and state governments in order to create an industrial extension network. The ARPA/TRP announcement required that federal funds be matched at the 100 percent level or higher—in most cases by the state government. This creates the possibility that the states that have funds for industrial extension will get the federal monies. The states are already smarting under the shifting of the tax load from federal to state levels that occurred during the two previous administrations. In addition, a state's economic development agenda may be at odds with the federal program's objectives. For example, a state's primary focus may be to entice U.S. companies to move there from another state—a zero-sum game from a federal perspective. To develop a national system of industrial extension agencies, we may also have to develop a cost-sharing partnership between state and federal governments.

A national electronic network that will serve manufacturing with the information needed for technology (industrial extension) and for commerce (virtual enterprise) has been proposed by both public and private agencies. The federal government has viewed the funding of this network as an acceptable role in developing the infrastructure needed by agile manufacturing in the future. Most of the emphasis has been on the electronic highway and on communication protocols or standards, with very little discussion about the transactions that will flow on this network. There will be two customer groups on this network: the extension agents and the companies. The communication between companies may flow through the extension agents initially, rather than direct company-to-company communication, due to the reluctance of the small manufacturer to make direct use of such capabilities.[16] The extension agent will use the network to post problems for other agents, to access national databases, and to receive professional training. The range of proposals to establish such a network, the identification of users, and the ability to demonstrate value are a few of the issues that must be resolved.

Another major issue is the relationship between a subsidized government service and private sector providers of the same services. The general argument is that the front-end services for small manufacturers, such as problem identification, should be government-subsidized, while the downstream services, such as in-house improvement projects, should be paid for by the firm at the fair market value. But there remains the difficulty of defining the boundary between these two types of services as well as the responsibility for training and technical support.

Whereas initial state efforts in industrial extension were based in the public universities and community colleges, there has been a shift to public not-for-profit economic development organizations as the source of these services. The university is expected to provide research and technology development, while the community college has been the focus of training programs. (Chapter 8 discusses this shift in the role of educational institutions.)

Finally, there is the question of who should be served. The current focus is on the 360,000 smaller manufacturing establishments, with little said about ownership (foreign versus domestic), products, or customers. We do not have the resources to serve this entire population; it is estimated that this would immediately require hiring over 2,000 extension agents. How, therefore, will the firms be selected for assistance?

A Brief History of the Federal Role in Industrial Extension

Industrial extension has existed on the state level for over three decades, usually managed through the state university system. It was originally modeled after the Agricultural Cooperative Extension Service, which started in 1914 and uses field agents to communicate new technologies and products directly to farmers. In 1987, the total funding for the Extension Service was $1.1 billion ($339 million from the federal government and $801 million from state and local sources), and the service employed 16,000 people. In addition, the federal government spent $822 million for agricultural research at its 148 research stations.[17] There is no question that agricultural extension has been highly successful in improving the productivity of U.S. agriculture; it is equally evident that many years of experience and very large investments were required to achieve this success. By comparison, U.S. efforts at industrial extension have been scattered, limited, and of short duration.

The Industrial Extension Service at Georgia Tech (the largest program of its type) was founded in 1956. The federal government first entered this arena in 1964, using limited federal funding from the State Technical Services (STS) Act to disseminate technical information to manufacturers through programs operated by the individual states. Non-university programs began to appear, usually operated by state economic development agencies through non-profit independent corporations. The PennTAP program at Pennsylvania State University and the New York State Science and

Technology Foundation were founded in 1965 under this program. Funding for STS was far too limited for its mission, and the program was ended in 1969 after a review of its benefits concluded that it was not successful.[18]

Development of industrial extension programs did not pick up again until the early 1980s after the relative industrial decline was a well-established trend in the United States and job loss in manufacturing had become a political issue on the state level. A recent study showed that 84 percent of the 42 state industrial extension programs with a technology focus were established since 1980.[19] It was during this period that the role of the small manufacturer as the infrastructure of U.S. manufacturing was recognized and became the focus of many of the new programs. The federal government again entered this arena in 1988 with the Omnibus Trade and Competitiveness Act, which established the Manufacturing Technology Centers (commonly called "Hollings Centers") program. The designers of the MTC program argued that the solution to improving the competitiveness of smaller manufacturers was to help them adopt advanced manufacturing technologies.

In addition to the MTC program, the 1988 Omnibus Trade and Competitiveness Act established two other complementary programs: the Advanced Technology Program (ATP) and the State Technology Extension Program (STEP). ATP provides funds to assist U.S. businesses to carry out research and development on "pre-competitive, generic" technologies. The program was funded at $48 million in 1992, and $68 million in 1993. The Clinton administration has proposed to double the funding for this program in 1994 based on the view that the multi-year, multi-million-dollar industrial joint venture will be a real stimulus to the economy. (This "supply side" program is described in Chapter 3.) The STEP program, on the other hand, has been funded at a very low level (less than $1 million per year) and has concentrated on small grants to states to aid in the design of economic development plans. Under TRP, funding for STEP should be somewhat increased, based on the perceived need for such planning at the state level.

There exists, however, little formal analysis of industrial extension to indicate which are the most effective approaches to providing these services, the extent to which these services are already provided by the states or by large corporations for their suppliers, and whether these services have the desired impact on the small manufacturer, that is, whether they increase small manufacturers' productivity and competitiveness, resulting in an improvement of the manufacturing sector of the economy. The 1990s will see two definite trends. First, large manufacturers will continue to downsize and increase their technological dependence on their smaller suppliers. Second, job growth will occur primarily in smaller firms. With job creation a major objective of the Clinton administration, the spotlight will remain on the small to medium-sized manufacturing firm as a source of these new jobs, and on industrial extension coupled with training and tax incentives as the public policy mechanisms to achieve this goal.

A comprehensive study of industrial extension, performed in 1989 and published in June 1990 by NIST, presents a picture of a wide variety of overlapping and competing programs on the state and federal levels with diverse services and constituencies.[20] This view was reinforced by a second study published in 1991, which focused on the 42 programs in 28 states that specialized in technology assistance.[21] The proposed infusion of federal money into this "system" will be both wasteful and counterproductive unless a national plan with national goals is first established. This plan must clearly delineate the state and federal roles in industrial extension and establish incentives for the states to modify their programs accordingly.

The Rationale for Industrial Extension

The justification for the proposed expansion of state and federal industrial extension activity is based on a number of premises which are reviewed in detail in the next section. We have found that these premises are generally sound and that there is a rationale for government participation in industrial extension. They are:

•Smaller manufacturers provide the infrastructure (supplier base) of U.S. industry; improving their competitive capabilities is essential for the expansion of industrial activity.

•Many smaller manufacturers are deficient in modern manufacturing technology and practices, lagging their offshore counterparts, thus limiting their productivity and their contribution to the infrastructure.

•Improving the competitive capability of the smaller manufacturers is a function of their ability to increase their productivity through the adoption of advanced manufacturing technologies and practices.

•Industrial extension services are able to transfer advanced manufacturing technologies to smaller manufacturers and assist them in implementing permanent improvements to their processing capabilities.

Underlying these premises is the proposition that publicly funded industrial extension services operate for the public good and provide a long-term economic benefit to their service regions. In addition, these premises have an important driver which constitutes a fifth premise:

•Smaller manufacturers are increasingly being called on to participate in the design and redesign of their customers' products, that is, the role of the smaller manufacturer has changed from "make-to-print," to the sale of their knowledge or core competence.[22]

Before evaluating the federal and state policies that respond to these premises, we must first examine the validity of the premises and the need for expanding industrial extension. We concentrate on industrial extension as a delivery mechanism for manufacturing technology and then define a strategy to delineate state, federal, and private sector roles and responsibilities in a greatly expanded industrial extension program.

An Examination of the Premises

•Smaller manufacturers are the infrastructure of U.S. industry.

While this premise is valid (see below), the number of firms that are in a position to contribute to the industrial economy either through value-added production or as suppliers to the end item producers is far less than the 360,000 figure quoted in federal legislation for the past five years. First, this figure is the number of establishments, not the number of firms.[23] Second, there are four characteristics that differentiate these establishments: ownership, customers or market, size, and industry. We estimate, for reasons explained below, that the true number of firms that could benefit from industrial extension and pass its benefits on to their customers, and hence to the national economy, is closer to 50,000 than 360,000. The design of an industrial extension delivery network should take these factors into account to focus public funding into the subset with the highest potential. Unfortunately, it will create "winners and losers."

•Smaller manufacturers are deficient in modern manufacturing technology.

This premise is true, but we must recognize that non-technology issues, such as marketing, finance, and regulatory compliance, may play a greater role in a small firm's strategic planning.

Two studies indicate that the rate of technology adoption was a function of ownership and customer, with foreign-owned firms and DoD suppliers dominant.[24] Shapira's analysis indicated that U.S. firms were far behind Japanese firms in technology adoption.[25] During the past four years, a large number of studies have been conducted to define the technology needs of the smaller manufacturer. Although some were merely inventories of what companies did or did not have,[26] others addressed the projected acquisition of new technologies by smaller manufacturers.[27] The consensus was that the smaller manufacturer had a low rate of technology adoption but planned to acquire programmable technology during the next 3–5 years.

Broader studies of the needs of smaller manufacturers have shown that concerns over government regulation (especially environmental regulation) and the need for business assistance (marketing and finance) generally took precedence over technology needs.[28] State programs designed to assist smaller manufacturers, however, have concentrated on technology extension. According to Clark's study, 40 states have technology assistance programs, while only 30 states have business assistance programs and only 31 states have financial assistance programs.[29]

The smaller the manufacturer, the more likely it will lack technological self-sufficiency. Consequently, the decision to acquire software (CAD, cost estimating, CAE, CAM, etc.) is extremely difficult for the smaller manufacturer—more difficult than equipment acquisition where the cost of acquisition and the savings payback may be more readily estimated.

•Improvement is a function of the adoption of advanced manufacturing technologies.

As stated above, non-technology issues may have greater weight on the competitive ability of smaller firms. While direct evidence of the relationship between technology level and productivity is lacking, there is some evidence from foreign programs that this is true.

Other countries have established programs to provide technical assistance to improve the competitiveness of their smaller manufacturers.[30] These include the Emilia-Romagna program in northern Italy, Baden-Wuerttemberg in West Germany, and SABRAE in Brazil. It should be noted that these programs include assistance in business functions (accounting, finance, marketing, export) as well as technical support thus allowing the entire firm to be assisted. This "total" approach recognizes the interrelationship between business and technical factors in the makeup of smaller firms. Recent studies in Indonesia[31] and the former Soviet Union[32] indicate that the lack of technically competent smaller manufacturers is the primary factor retarding the growth of the manufacturing industry for export.

•Industrial extension services are able to transfer advanced manufacturing technologies.

There has been substantial debate on the appropriate methods of measuring the impact of extension services on client firms. Evaluation of existing technology transfer programs focuses on case studies, because it is very difficult to link technology transfer actions to common success measures such as increases in market share, sales, or profit, due to other influences (both internal and external) on the firm.[33] State and federal funders, however, are placing additional demands on technology extension agencies to develop and present more general measures of performance, including such measures as the increase (or preservation) of jobs, increase in tax revenue, increase in exports, etc. Combined with the complication of politically-defined issues, such as defense conversion and the "dual use" strategy, this creates a very difficult problem in evaluation. Current funding strategies, for example, give priority to providing federal support for firms that are converting from the defense to the civilian market with special emphasis on firms whose product can be sold in both markets, that is, "dual use" products.

NIST has initiated studies in the area of regional impact to measure macroeconomic performance, but little can be done to deal with the central issues—demonstrating cause-effect relationships and measuring the effect of the time lag between the technology transfer action and the improved performance result. The use of benchmarks to compare a firm to "world class" firms appears to be the direction that evaluators have taken, and professional benchmarking organizations have made their appearance. Benchmarking requires the ability to measure base-line conditions within the firm before the assistance is provided and then periodically (e.g., every six months) afterwards. The cause-effect relationship, however, is still handled anecdotally.

An unpublished survey by Joe Paterno, director of the Industrial Research Center at the University of New Hampshire, covered 38 programs in 25 states, and estimated that the states were spending an average of $5,000 to either retain or create a job, and that the state tax generated averaged more than one dollar for each dollar spent.[34] Even though the data was incomplete, this survey is an indicator of the value of extension services to the state economy.

•Smaller manufacturers participate in the design and redesign of their customers' products.

The documentation supporting this premise comes from two sources. First, from the designers of the manufacturing system of the future who have defined the "virtual corporation"[35] and the "seamless enterprise."[36] Both concepts are based on a partnership between a manufacturing firm and its suppliers which includes excellence, trust, opportunism, and the lack of traditional boundaries. Larger manufacturers have embraced these concepts and are in various stages of implementing their external relations aspects through supplier-development programs.

The second source of documentation is the information technologists who are supporting the development of electronic networks to link cooperating firms. This development takes two forms: one-on-one electronic exchange of information between firms, and the electronic utility that allows a user to access a variety of databases and to communicate with other users on the network. Electronic networking is emerging as an infrastructure priority under the Clinton administration. Electronic network utilities such as EINet and Factory Net America are in the planning stage, and existing operations such as Internet have attracted major interest. In addition, communication protocols such as PDES/STEP under the DoD CALS initiative are being developed. The need for a standardized system for the transfer of technical information has become a major building block of the industrial revival. (See Chapter 5.)

We may conclude from the foregoing that the assumptions upon which the concept of the need for industrial extension is based are reasonable, and that there is a need to expand industrial extension into a national system that supports and enhances the high impact group that we have identified.

The Infrastructure Role of Smaller Manufacturers

There is widespread acceptance of the concept that the smaller manufacturers form an industrial infrastructure essential to the health of the manufacturing sector of our economy. This was put forward formally by Senator Hollings in the 1988 legislation that established the Manufacturing Technology Center program and has been reiterated in subsequent legislation such as the proposed National Competitiveness Act of 1993[37] and the DoD Technology Extension Program.[38] In a white paper entitled "Manufacturing Excellence Partnership," written for the U.S. Department of Commerce's Technology Administration, NIST restated the argument that "America's small and medium-sized manufacturers are crucial to our international competitiveness and economic vitality."[39]

Japanese officials take this assertion seriously, as evidenced by the fact that the Kohsetsishi, Japan's industrial extension organizations, through 170 centers funded at $500 million per year, provide direct manufacturing assistance to Japan's smaller manufacturers.[40] Shapira credits the "broad robust base of small manufacturers" as a key factor in Japan's success. These firms have become the source of high quality components for Japan's OEMs (large or Original Equipment Manufacturers) and have, with the assistance of the Kohsetsishi, become innovators themselves. There is, therefore, a dual role for the smaller manufacturer—as supplier to OEMs and as a developer of new technology through a "core competence."[41]

If we view the role of the smaller manufacturers as suppliers to end-item producers, there is some data to support the infrastructure argument. During the last 15 years, the larger firms have downsized their manufacturing operations to focus on design and marketing, by outsourcing component fabrication to smaller manufacturers. In the 1980s, employment in manufacturing dropped by 10.6 percent.[42] Firms with more than 500 employees reduced employment by 18.9 percent due to a combination of poor business conditions, loss of market to imports, increasing productivity and the trend toward outsourcing. On the other hand, employment remained constant in the aggregate of firms employing less than 100 people.

Of the 360,000 manufacturing establishments in the United States, 90 percent have fewer than 100 employees. Understanding the relationship between establishment statistics and enterprise or firm statistics requires an examination of ownership.[43] The 1990 Census of Manufacturers indicated that 1.6 percent of these establishments were foreign-owned. Norsworthy and Jang's study of technology adoption,[44] which combined the 1987 Census of Manufacturers' study with the 1988 study of manufacturing technology,[45] indicated that foreign-owned firms had adopted technology at a rate that was 1.5 times higher than U.S.-owned firms in the same standard industrial classifications (SICs).

Smaller firms account for 50 percent of the manufacturing employment and 60 percent of the discrete parts manufactured in the United States.[46] A study of the metalworking industry in Michigan indicated that two-thirds of the 1800 firms were suppliers to the auto industry.[47] We estimate that over 50 percent of smaller manufacturers are in a supplier relationship to larger manufacturers.

Economists, geographers, and economic development analysts use the concept of agglomeration to describe a set of multiplier effects (which are not always positive) that accrue to a concentration of particular activities in particular places.[48] It is the interaction of these firms in buying and selling from each other as well as cooperating in product design and sharing process knowledge that produces a substantial impact on the regional economy. Two recent studies have taken the approach of developing agglomerations of firms to be served by extension services.[49] The firms are selected for agglomeration on three bases: Standard Industrial Classification (SIC), value added, and supplier role. Norsworthy and Jang's re-examination of the 1988 study's database of over 10,000 establishments in SICs 34 through 38 also focused on subsets that were primary contributors to either the supplier base or GDP.[50]

The 360,000 manufacturing establishments that might be served by industrial extension services are by no means equal in their need for assistance, their ability to absorb it, and the benefits to the economy that might flow from such assistance. The cost of bringing manufacturing extension services to all 360,000 could be prohibitive. Services should be designed for a much smaller target group, at least initially, in the hope of gaining measurable evidence that government investment in such services is economically justified. What might be the criteria for selecting the most appropriate recipients for extension services?

NIST has already faced this question in searching for a geographic criterion for locating Manufacturing Outreach Centers (MOCs). Fogarty and Lee's study of agglomeration was used as the basis for the NIST mapping of industrial clusters which may be used in selecting candidate organizations for extension services.[51]

From the foregoing, three characteristics emerge as important delineators for differentiating manufacturers that have high or low potential to improve and affect the economy:

•Size: There is a substantial difference between the capabilities of very small firms (fewer than 20 employees) and small firms (20–100 employees). The Northeast Manufacturing Technology Center (NEMTC) has found that very small firms generally lack the employee skills and the capital resources necessary to adopt advanced technologies.[52] They tend to be in a startup mode and rely on infusions of venture capital or Small Business Innovation Research grants.[53] The very small firm is in the process of adopting low-end technologies such as computer-aided design (CAD), while the small firm is trying to link its CAD system to computer numerically controlled (CNC) machines through computer-aided engineering software. Small to midrange firms (20–499 employees) account for 116,000, or 32 percent, of all establishments. They are more likely to be able to absorb technical assistance than very small firms.

•Business Sector: The agglomeration approach indicated that firms in the SIC categories 22 (textiles), 23 (apparel), 25 (furniture), 30 (plastic products), 33 (primary metal), 34 (fabricated metal), 35 (machinery), 36 (electronic equipment), 37 (transportation equipment), and 38 (instruments) have the highest potential to contribute to the economy from the perspectives of value added and their role as suppliers to larger firms. These firms account for 187,000 establishments, or 52 percent of total manufacturing establishments. In addition, 68,000 firms with 20 to 499 employees are in these SICs, and offer an attractive candidate class for manufacturing extension services.

•Customers: The role of this subset of firms as suppliers to larger firms or OEMs has been established by a number of studies as well as the service experience of NEMTC.[54] Two studies, however, indicate that firms selling directly to the Department of Defense or in a lower-tier supplier relationship with a firm selling to the DoD are much more likely to be technically advanced than their non-DoD counterparts.[55] However, studies of these firms to determine the impact of defense conversion indicate that these firms need more assistance in marketing and finance than in the technical improvement of their manufacturing operations.

Taking size, business sector, and customers into account, the highest impact on the national economy should be achieved by assisting firms that are in the 20 to 500 employee size range, in the above-defined SIC set, and are suppliers to larger original equipment manufacturers. In addition, separate studies have shown that defense contractors and foreign-owned firms are technically superior. These firms should be included in the high impact group because they have a skill base to build on, and it would not be politically desirable to exclude them.

This subset (as defined by size, customer and SIC) is approximately 50,000 firms, out of 360,000 establishments. An alternate approach to defining this subset is to determine the service capability of the 130 to 170 industrial extension organizations projected by the ARPA program. This is estimated at 60,000 to 80,000 firms. A National Academy of Engineering study sets this goal even lower at only 20 to 25 percent of the 360,000 establishments.[56] We previously stated that the MTCs assisted 1200 firms during their first three years of operation. The impact of their operations would have been increased by focusing on firms with great potential to affect the economy. This conclusion is, however, the result of an examination of existing studies and data, which are fragmented and limited in scope. Before a major fund commitment is made in the area of industrial extension, the federal government should examine these premises and, if found valid, locate federal assistance where these firms agglomerate geographically. It should be noted that NIST has developed such a map and is considering its use in the selection of MOC sites.

It would be both difficult and politically undesirable to exclude any firm that requests services, or to create eligibility requirements that were unenforceable. New York State's Industrial Effectiveness Program used to exclude any firm whose main competition was also a New York State firm, until it realized that it was simpler to offer assistance to both firms. The fact remains, however, that small firms usually do not actively seek help unless initially contacted. (This supports the need for outreach programs to encourage the use of extension services.) If we couple this behavior with a selection policy for service organizations that is based on geographic or industry clusters, or a combination, we can target most resources toward the high potential firms.

Sources of Technical Assistance

To develop a strategy for expanded technology extension operations, we must first list the primary sources for the transfer of
advanced technology. These sources can provide physical products, documentation, or individual expertise. This section focuses
on state and local government providers of technology extension
services to manufacturing businesses; the next section describes
the different kinds of services. In addition to these state-based
services, which are the most significant and which will be discussed
in detail in the next section, there are at least eleven other sources
of technology help available to small firms.

• Consultants and other independent experts

• Commercial vendors of tools, processes, and software

• University, college, and two-year college research and demonstration facilities

• State, foundation, and federally funded research centers whether
university-based or independent

• Supplier development and TQM (total quality management)
programs being promoted by such large industrial firms as Motorola,
Xerox, IBM, and Procter and Gamble

• The seven current MTC operations (which are a mix of service and
source) as a source of experts

• Industrial associations and research organizations such as
TCSquare (Textile and Clothing Technology Center)

• Commercial databases and collections of experts such as Teltech
and Best

• Existing technical information operations in DoD, NASA, EPA,
and other federal agencies

• NTIS as a knowledge access source to the federal laboratories'
publications. (It is doubtful that access to the bench scientists
within the federal laboratory system can be increased.)

• The federal laboratories as a source of products which, under the
Stevenson-Wydler Act, should be transferred to public use or
commercialized. (For a more detailed discussion, see Chapter 4.)

Not all of these resources are readily available to smaller manufacturers. For example, private consultants, who cannot afford to serve the assessment and education function for the small business client, defer to the state or federal extension agency to define the problem and develop the scope of services. Consultants can perform profitable services once the market for their services is established.[57]

This range of sources helps emphasize the need to educate smaller manufacturers to make an intelligent choice and, once that choice is made, provide training and support for smaller manufacturers to implement their choice. This defines the role of technology extension as a service or transfer organization, but does not solve the problem of linking them to the sources. This is addressed in the industrial extension model, below.

A Taxonomy of Industrial Extension Programs and Models

Forty states currently have some form of technology extension program. NIST's 1990 study of these programs indicated that they fell into seven categories, distinguished by their institutional structure:[58]

•Business Assistance: general business management information including personnel, accounting, and legal services

•Incubators: office and lab space for startup companies, including shared business services

•Research Parks: planned groupings of technology companies, usually university-affiliated

•Seed Capital: research grant and product development grants to projects in early stages of development

•Technology Assistance: information exchange and outreach programs that assist in application of new technologies

•Technology/Research Centers: university-based, industry affiliated research centers

Shapira's 1992 study defined four types of technology extension programs, distinguished by their technical depth:[59]

•Full Extension: based on agricultural extension; includes research, development, and deployment, with most research university-based

•Modified Extension: similar to full extension but lacks research; focus is on technology deployment through field agents

•Center/Satellite/Gateway: small network of regional state centers offering experts in specialized areas; intake is accomplished through separate service offices. This is a three-tier system with services provided at the second satellite level.

•Brokering and Networking: large number of localized industrial clusters, aided by network brokers and public and private technical experts

The recently announced ARPA Technology Reinvestment Project defines three forms of Manufacturing Outreach Centers, distinguished by their relationship to the Manufacturing Technology Centers:[60]

•A smaller version of an MTC acting as a satellite to that MTC

•A free-standing entity serving a natural service region (geographic, industry, or both)

•A seed for a potential MTC

Our experience indicates that no single model can satisfy the diverse regional needs of smaller manufacturers. The choice of model must be based on whether it will focus on a strategic industry or on a technology need in its service region, and its "fit" with existing programs in the region.

NIST's Programs

The Manufacturing Technology Centers were established in 1989 under the Omnibus Trade and Competitiveness Act of 1988. There are currently seven regional centers (three established in 1989, two in 1991, and two in 1992) engaged in transferring advanced off-the-shelf manufacturing technology to smaller manufacturers to improve their productivity and hence their competitiveness. The MTCs vary in focus, including some with a narrow regional focus (for example, Great Lakes MTC in Cleveland), some with a broad regional focus (Northeast MTC in New York), and some with an industry focus (Midwest MTC at the Industrial Technology Institute in Michigan). Five of the seven are linked to existing programs in the states they serve. Evaluations of these programs conclude that five key lessons were learned:[61]

•Smaller manufacturers needed appropriate technology, not necessarily state-of-the-art.

•Services should be driven by industry needs.

•Programs must reflect the diversity of states and regions.

•Technology assistance cannot be provided independent of financial, management, and human resource activities.

•There is a need to establish smaller assistance centers in regions with lower industrial density.

In addition, NIST is in the process of establishing linkages to complementary programs located in other federal departments. For example, with the U.S. Department of Agriculture, NIST established a demonstration project in Kansas to use cooperative extension agents as industrial extension agents in rural areas. Other demonstration projects are underway with the Small Business Administration (financing), the Department of Labor (human resource development), the Department of Education (training), the Department of Energy (technical expertise), and the Department of Defense (defense conversion).

The State Technology Extension Program (STEP) at NIST has made small grants to individual states to assist in the development of state-wide economic development programs. In 1993, STEP shifted its focus to regional economic development programs encouraging multi-state cooperation.

NIST's current plans include expanding the MTC program (to 30 MTCs by 1999), establishing a smaller version of the MTCs called Manufacturing Outreach Centers (the goal is 100 MOCs by 1999), and expanding the STEP program to 40 states by 2005.[62] Similar concepts have been put forward in President Clinton's platform and in the FY93 Defense Authorization and Appropriation Bills. We conclude that there will be a major expansion of industrial technology extension activity starting in 1993, funded, most likely, under the rubric of defense conversion and administered jointly by the Departments of Defense, Commerce, and Energy.

This represents a major departure from the Reagan-Bush approach of providing tax incentives to encourage the adoption of advanced manufacturing technology. Tax incentives cannot by themselves overcome the natural resistance of smaller manufacturers to adopting software-based technologies. They can, however, complement technology extension efforts as part of a broad-based set of alternatives and assistance programs available from local, state, and federal agencies.

Implementation Issues

In defining a federal strategy, several issues should be addressed. They include the need to leverage public funds, that is, to maximize the number of companies served while minimizing federal funding through state match, cost sharing, and industry participation. Similar programs in Japan and Italy provide five to six extension agents per 1000 companies served. The best-funded state programs in the United States provide only one agent per 1000 companies. An answer would be to focus industrial extension nationally on the subset of about 50,000 target firms previously defined that offer the highest impact on the economy. This would bring services for the target firms to the same level as in Japan, and could be accomplished through a selection and placement process for industrial extension services that gives priority to the areas of highest agglomeration.

State programs are highly influenced by state budget policies. As a result, we have witnessed substantial cutbacks and even elimination of state extension programs at a time when the economic need for these programs is highest. In the future, such programs will need trust funds or a dedicated revenue stream to insure continuity. A state program, however, cannot spend money outside its state and must establish economic development goals specific to its state. This limitation restricts the state-funded program, unless federal funds are made available. Federal funds should, therefore, be used to leverage successful state operations and to link them for mutual support.

Networking in all forms (supplier, quality, trade, industry, regional, etc.) has become the primary approach to leveraging the limited state and federal funds. While European models have been studied and described, there is limited documentation of U.S. efforts in networking.[63] Approaches and models should be defined and codified. (See Chapter 5.)

Advanced manufacturing technology is heavily dependent on information technology—communication, access to databases, access to tools, and access to other companies and services. The need for standards in this area has prompted the opening of the DoD's CALS program to civilian use with the dissemination of PDES/STEP (Product Data Exchange using Step/Standard for Exchange of Product Data). Network proposals including the expansion of Internet and EINET are currently under review. But this only supplies two of the three elements of an "information highway"—the network and the standards. The technology extension program provides a natural source for defining the third element—the transactions that will take place on these networks. The logical agency to deal with this issue is NIST, because of the infrastructure and standards required.

The necessary training component of industrial extension, as well as the proposed national apprenticeship program, depends on the community college "system" as the principal resource. But the community colleges are not really a system; rather they are a collection of state-level networks and individual fiefdoms linked by state funding and regulation. This resource must be examined to determine the optimum organization required for the desired services before investments in apprenticeship programs are made. Training must be focused regionally within a state based on the needs of the region's employers. Therefore, training should be integrated with industrial extension in a state-wide economic development plan, which is in turn linked (via STEP) to a national plan.

The appropriate interface between state and federal programs has not been defined; without doing so, we cannot develop the appropriate service models. For example, the Northeast Manufacturing Technology Center (NEMTC) established a relationship with New York State's Industrial Technology Extension Service (ITES). NEMTC supplied technical services such as field projects and demonstrations, while the ITES field agents provided outreach (marketing of these services) to client firms. The ITES field agents, however, had the responsibility to differentiate between the business needs of the firm (which were referred to existing state programs such as the Industrial Effectiveness Program) and the technology needs (which were referred to NEMTC). This was facilitated by hiring highly experienced people with business and engineering backgrounds as field agents. This is a successful relationship, but can it be emulated by other states?

These issues have been taken into consideration in our design for a federal strategy for industrial extension. We have further defined the intermediate-term goal of industrial extension to be the creation of jobs through expanded sales and markets, even though the immediate effect of a productivity improvement is reduced employment. This creates the need to provide a capital fund to facilitate the growth of the smaller firm after improved productivity has increased its sales. Thus, industrial extension becomes linked to marketing, training, and financing.

Drivers and Blockers

Before defining a national system for industrial extension, we should look at why the smaller manufacturers will or will not adopt advanced technologies on their own. That is, what are the drivers of technological modernization and the blockers that inhibit it?

The demands of a smaller manufacturer's customers are the principal drivers that motivate the smaller manufacturer to adopt advanced technology. If smaller manufacturers do not respond quickly to the requirements placed on them by their customers, they will lose business. However, while past demands, such as just-in-time, could be handled by solutions external to the production system (for example, a firm could rent a warehouse and fill it with finished goods), the requirement to participate in the customer's design and development process requires substantial internal change. In addition, these changes affect the areas that constitute the greatest blockers, which include a lack of requisite skills, core competence or knowledge, and the capital necessary to acquire advanced technologies.

Smaller manufacturers are in transition from vendor, to supplier, to "partnering" with their customers to reduce product development cycle time through participation in the design and development process that is software driven. This role may require that participants acquire programmable equipment and communication systems. Software acquisition is the most difficult decision associated with the blockers defined above.

Smaller manufacturers have traditionally depended on the salesperson and the consultant to provide advice on the acquisition of advanced technology. This support has generally been effective in equipment acquisition, where the ratio of capital investment to training is 10:1, but has been ineffective in software acquisition, where this ratio can easily be 1:10. A smaller manufacturer is highly dependent on the technology vendor for technical support. Therefore, there is a tendency for smaller manufacturers to seek off-the-shelf, proven technologies.

The list of blockers is quite impressive. Shapira divided them into four classifications:[64]

•Firm-level: including poor use of existing technology, insufficient training, lack of information, lack of expertise, lack of time, no long-term strategy, inadequate finance, and no cost justification

•Business infrastructure: short-term customer relationships, poor advice and service from technology vendors, consultants too narrow, trade associations not technology-focused, weak linkages with other firms in same industry

•Social infrastructure: low-quality education and training; existing modernization programs inaccessible, lack competence, or are not technology focused; universities are research-focused

•National policies and attitudes: uncertain attitude toward manufacturing in the United States; federal emphasis on R&D and advanced technologies; unhelpful macro-economic, tax, and trade policies

It is unlikely, given the character of the blockers, that smaller manufacturers will adopt advanced technology on their own, despite indirect drivers such as tax incentives, loan programs, enterprise zones, industrial bonding, and training programs. They need the support and technical assistance of an outside agency, whether it be a consultant or a publicly funded technology extension organization. The indirect drivers, however, especially in the area of financial and training assistance, can be effectively used in conjunction with technology extension.

Federal Strategy for the 1990s

From a policy perspective, the federal government is now committed to assisting smaller manufacturers, the industrial base of the United States. The federal government can play an important facilitating role in the industrial revitalization of the United States by pursuing these long-range goals:

•Focusing on the states as the primary service provider by extending support to develop and enhance state-based industrial extension programs with regional missions and priorities

•Collecting and defining sources of technology and knowledge and, in the cases where such knowledge does not exist, developing new sources of support

•Improving the delivery capability of the programs by linking them to sources of technology and knowledge through an electronic network with the appropriate interface standards

•Financing research on manufacturing extension in both the policy and operations areas

•Evaluating the overall and relative effectiveness of the delivery models and adjusting their support of the operational units to improve the performance of the system

The Industrial Extension Model

The appropriate model is a two-tier system for industrial extension consisting of a set of state programs supported by a federal information network. This should be developed and partially funded by the federal government. The fundamental building block of this system will be the Manufacturing Outreach Center (MOC), defined at a NIST-sponsored workshop conducted in December 1992. The participants in that workshop concluded that the important characteristics for an MOC are:

•The organization should have a proven track record with a community focus which includes both participation in the community and the confidence of the community.

•The services provided by this organization should be a function of the needs of the client base that are not provided by the other service providers in the community, i.e., they should "fit" with the other providers.

•The outreach and field services must be provided to allow direct contact with and continued support for the client firm.

•The services should focus on the application of best practices for the continuous improvement of the client firm's productivity.

•The organization should have capabilities in benchmarking and selection of best practices.

•The services must be flexible, i.e., be able to respond to such issues as ISO 9000 (the European Community/Common Market quality standard) and environmental compliance.

•The organization must be able to treat the entire firm (which may comprise geographically distributed establishments) whether the problems are technical, business related, quality, or human resources related.

The MOC model, with its focus of technology transfer and service delivery, should be used as the basis for enhancing state-level industrial extension efforts. Selection criteria for the new MOCs would probably include analysis of defense conversion impact in the proposed service region. The MOC or state role is clearly service-oriented and involves intake, problem definition, solution definition, and solution implementation. The MOC must be able to treat the whole firm and assist the firm in the resolution of technical, marketing, financial, and regulatory problems.

The MOC should be small—five or six professionals serving a region of 800 to 1000 manufacturing firms. It is unlikely that an organization of this size would have the range of expertise to treat the whole firm and, therefore, a support network providing access to more specialized experts and services is required. The role of the federal government should be to establish this support network.

A federally sponsored organization would have to be established or contracted to develop cooperative agreements with existing source providers and make these services, knowledge, and experts available to the MOCs through a national electronic network, such as an enhanced Internet. The first phase of development should focus on meeting the needs of the MOCs' field agents in such areas as:

•Communication (e-mail) and bulletin board capability—allowing the MOC field agent to "post" a problem and seek assistance from field agents in other MOCs

•Access to databases and experts—including university manufacturing research centers, commercial databases, consultants, etc.

•Access to "tools"—including questionnaires, checklists, expert systems, and documentation that would directly aid the field agent in providing assistance to a small firm in such areas as total assessment, ISO 9000, Total Quality Management, CAD selection, etc.

During the second phase of development, client-level services should be added, such as access to the Federal Register, subcontractor lists, used equipment, state contracts, transfer of drawings and product specifications, etc. (similar to the Massachusetts-based TECnet operation[65]). These services will provide data standards and communication protocols to insure that transactions take place in a uniform and accurate manner. PDES/STEP offers this capability and the entire network program can be used to demonstrate the viability and effectiveness of CALS. (See Chapter 5).

In addition to coordinating technology and knowledge sources for the network, the federally sponsored organization should also provide services through the existing MTCs or contract for services in such areas as:

•Training extension agents—providing a skill set to the variety of people who will be hired as field agents during the expansion of the industrial extension program

•Designing and developing "tools" for the extension operations, as described above.

•Evaluating the effectiveness of the state-level programs

•Coordinating with and establishing agreements with other federal service and knowledge providers including the Departments of Labor, Agriculture, and Energy, the Small Business Administration, etc.

The success of this two-tier structure will be based on its ability to recruit existing state level organizations with proven capabilities into a cooperative network-linked group of service providers. The federal initiative will have to provide 30 to 50 percent of the funding to obtain the cooperation of the state programs.

Conclusion

This chapter presented a rationale for technology extension and a proposed strategy for implementing a national system based on federal sources of knowledge and assistance and state agencies for the delivery of services. It also identified issues for which more information is needed, such as evaluation, networking, and funding. Evaluation, especially, will require a collaborative effort between the state providers and the federally funded providers, due to the shift under the Clinton administration from a productivity improvement focus for industrial extension to one of job creation.

The federal role should be to convene the states to work toward a system of delivery and support for manufacturers, in which the states are linked through a knowledge network that is directly funded by the federal government. Joint funding of the operational units by the federal and state governments through cooperative agreements offers the best chance of ensuring the effectiveness of delivery. Direct grants would only expand an already heterogeneous collection of services, without regard to identifying the most effective methods to improve the productivity of smaller manufacturers.

The federal role should not be to develop a new set of organizations, but rather to facilitate the conversion of the network of existing organizations into a national system with national goals. At the first annual staff conference of the Manufacturing Technology Centers, held in March 1993, Jack Russell, president of the Modernization Forum, summarized the challenge: "The debate now is not if a national system supporting industrial modernization is necessary, but how it should be built, at what final scale, by whom, and how quickly."[66]

Notes

1. National Institute of Standards and Technology, *Overcoming the Competitive Barriers Facing Smaller Manufacturers: An Analysis of the Need for the Manufacturing Extension Partnership* (Gaithersburg, Md.: U.S. Department of Commerce, No-

vember, 1992). Because manufacturing productivity rose slowly through much of this period, some part of the job loss resulted from that increase. Most of the job loss, however, came from imports and the recession.

2. Lewis M. Branscomb, "America's Emerging Technology Policy," *Minerva*, Vol. XXX, No. 3 (Autumn 1992), pp. 317-336.

3. Some of the studies that support this conclusion are: The Center for Strategic and International Studies, *Strengthening of America Commission: First Report* (Washington, D.C.: Center for Strategic and International Studies, 1992); Marianne K. Clark and Eric N. Dobson, *Increasing the Competitiveness of America's Manufacturers: A Review of State Industrial Extension Programs* (Washington, D.C.: Center for Policy Research, National Governor's Association, 1991); Modernization Forum, *Modernizing America's Industrial Base: Opportunities for Action* (Detroit: Conference Proceedings, May 14–16, 1991); National Academy of Engineering, *Foundations of World-Class Manufacturing Systems* (Washington, D.C., Symposium Papers, June 19, 1991); National Academy of Engineering, *The Challenge to Manufacturing: A Proposal for a National Forum* (Washington, D.C.: National Academy of Engineering, 1988); MIT Commission on Industrial Productivity; and Michael L. Dertouzos, et al., *Made in America: Regaining the Productive Edge* (Cambridge, Mass.: MIT Press, 1989).

4. Ernst & Young, *American Competitiveness Study: Characteristics of Success*, No. 58059 (Cleveland, OH: Ernst & Young, 1990); Office of Technology Assessment, *Making Things Better: Competing in Manufacturing* (Washington, D.C.: Congress of the United States, February, 1990); Philip Shapira, J. David Roessner, and Richard Barke, *Federal-State Collaboration in Industrial Modernization* (Atlanta: School of Public Policy, Georgia Institute of Technology, July, 1992).

5. Technology Administration, *Strategic View* (Washington, D.C.: U.S. Department of Commerce, November, 1991).

6. "Industrial Policy," *Business Week*, April 6, 1992, pp. 70–76.

7. Jeff Bingaman, Chairman, Senate Subcommittee on Defense Industry and Technology Committee for the Armed Forces, *FY93 Defense Authorization and Appropriation Bills* (Washington, D.C.: U.S. GPO, October 8, 1992).

8. William J. Clinton and Albert Gore, Jr., *Technology for America's Economic Growth: A New Direction to Build Economic Strength* (Washington, D.C.: The White House, February 22, 1993).

9. John A. Alic, Lewis M. Branscomb, Harvey Brooks, Ashton B. Carter, and Gerald L. Epstein, *Beyond Spinoff: Military and Commercial Technologies in a Changing World* (Boston: Harvard Business School Press, 1992), chap. 12.

10. Advanced Research Projects Agency, *Program Information Package for Defense Technology Conversion, Reinvestment, and Transition Assistance* (Washington, D.C: ARPA, March 10, 1993).

11. National Institute of Standards and Technology, Third Year Review Panel, *Manufacturing Technology Centers: Broad Programmatic Issues* (Gaithersburg, Md.: Report to the Secretary of Commerce, April 7, 1992), p. 20.

12. National Institute of Standards and Technology, Visiting Committee on Advanced Technology, *The Manufacturing Technology Centers Program* (Gaithersburg, Md.: Report to the Secretary of Commerce, October, 1990).

13. Robert E. Chapman, Marianne K. Clark, and Eric Dobson, *Technology-Based Economic Development: A Study of State and Federal Technical Extension Services*, Special Publication #786 (Washington, D.C.: NIST, June 1990), p. 28.

14. The roughly $3,000 per firm assisted represents the publicly subsidized portion of the extension service. After the initial subsidized service contract, most states require reimbursement for follow-on support.

15. NIST, Third Year Review Panel (April 7, 1992), p. 27.

16. Experiences with TECnet, a demonstration project in Massachusetts, and Teltech, a Minnesota-based information service, bear out this conclusion. The small firms preferred to work with the extension agent rather than make direct inquiries. In addition, they preferred documents rather than the assistance of the experts who were available under the service.

17. Chapman, Clark, and Dobson, *Technology-Based Economic Development*, p. 7.

18. Timothy P. Murphy, *Science, Geopolitics, and Federal Spending* (Lexington, Mass.: D.C. Heath, 1971).

19. Clark and Dobson, *Increasing the Competitiveness of America's Manufacturers*.

20. Chapman, Clark, and Dobson, *Technology-Based Economic Development*.

21. Clark and Dobson, *Increasing the Competitiveness of America's Manufacturers*.

22. Roger Nagel, *The Agile Opportunity* (Lehigh: 21st Century Manufacturing Enterprise Strategy, Iacocca Institute, Lehigh University, 1992).

23. An establishment is a single operating unit of a firm. A firm may consist of several establishments.

24. Maryellen R. Kelley and Todd A. Watkins, "The Defense Industrial Network: A Legacy of the Cold War," *Contractor Report* (Washington, D.C.: Office of Technology Assessment, U.S. Congress, November, 1992); J.R. Norsworthy and Show-Ling Jang, *Technology Adoption in U.S. Manufacturing: A Comparison of Foreign-Associated and Domestic Plants*, Working Paper 92-1 (Troy, N.Y.: Center for Science and Technology Policy, Rensselaer Polytechnic Institute, May 1992).

25. Shapira, Roessner, and Barke, *Federal-State Collaboration in Industrial Modernization*.

26. Department of Commerce, Bureau of the Census, *Manufacturing Technologies 1988: Current Industrial Reports* (Washington, D.C.: U.S. Government Printing Office, May 1989).

27. Georgia Tech Research Institute, *Georgia Manufacturers Speak Out: Results of the 1989 Manufacturers Survey* (Atlanta: Georgia Institute of Technology, 1990); Maryellen R. Kelley and Harvey Brooks, *The State of Computerized Automation in U.S. Manufacturing* (Cambridge, Mass.: Program in Technology, Public Policy,

and Human Development, John F. Kennedy School of Government, Harvard University, 1988).

28. Center for Strategic and International Studies, *Strengthening of America*; Modernization Forum, *Modernizing America's Industrial Base.*

29. Chapman, Clark, and Dobson, *Technology-Based Economic Development.*

30. Stuart A. Rosenfeld, "Regional Development, European Style," *Issues in Science and Technology*, Winter 1989-90.

31. Board on Science and Technology for International Development (BOSTID), National Research Council, *Rationale for Programs to Accelerate Growth in Indonesia's Manufacturing Industries* (Washington, D.C.: National Research Council, October 1, 1992).

32. William Martel and David Mussington, *The Problem of Defense Conversion in the Former Soviet Union*, Draft CSIA Working Paper (Cambridge, Mass.: Center for Science and International Affairs, Harvard University).

33. Among of the studies supporting this conclusion are: Peter J. Bearse, "Technology Transfer: Designing a State Strategy," *Commentary* (Washington, D.C.: Council for Urban Economic Development, Spring 1982); Robert Howard, "Can Small Business Help Countries Compete?" *Harvard Business Review* (November-December 1990); Office of Technology Assessment, *Making Things Better: Competing in Manufacturing* (Washington, D.C.: Congress of the United States, February, 1990); Philip Shapira, *Modernizing Manufacturing: New Policies to Build Industrial Extension Services* (Washington, D.C.: Economic Policy Institute, 1990); Philip Shapira, "Modern Times: Learning from State Initiatives in Industrial Extension and Technology Transfer," *Economic Development Quarterly*. Vol. 4, No. 3 (August 1990); Gene R. Simons, "The Experience of the Northeast Manufacturing Technology Center," *Proceedings of the Second Meeting on Modernizing America's Industrial Base* (Detroit, May, 1991).

34. Joseph Paterno, *A Survey of the Economic Impact of State Industrial Extension Programs* (Durham: New Hampshire Industrial Research Center, University of New Hampshire, February, 1993).

35. Nagel, *The Agile Opportunity*; "The Virtual Corporation," *Business Week*, February 8, 1993.

36. Dan Dimancesco, *The Seamless Enterprise: Making Cross-Functional Management Work* (New York: Harper Business, 1992).

37. H.R. 820 and S. 4, 103rd Cong., 1st Sess. (1993).

38. Les Aspin, Chairman, House Armed Services Committee, *A DoD Technology Extension Program* (Washington, D.C.: U.S. GPO, April 28, 1992); Bingaman, *FY93 Defense Authorization and Appropriation Bills.*

39. National Institute of Standards and Technology, *Manufacturing Excellence Partnership: A National Strategy for Manufacturing Excellence* (Gaithersburg, Md.: NIST, December 10, 1992).

40. Philip Shapira, "Lessons from Japan: Helping Small Manufacturers," *Issues in Science and Technology*, Spring 1992.

41. C. K. Prahalad and Gary Hamel, "The Core Competence of the Corporation," *Harvard Business Review*, May-June 1990.

42. Department of Commerce, Bureau of the Census, *Country Business Patterns—1990* (Washington, D.C.: U.S. GPO, 1991).

43. An establishment is a distinct business unit operating at a specific location. Most firms are very small and have only one establishment. The largest firms may have tens or even hundreds of them.

44. Norsworthy and Jang, *Technology Adoption in U.S. Manufacturing*.

45. Bureau of the Census, *Manufacturing Technologies 1988*.

46. Michael L. Dertouzos, et al., and the MIT Commission on Industrial Policy, *Made in America: Regaining the Productive Edge* (Cambridge, Mass.: MIT Press, 1989).

47. Industrial Technology Institute, Vol. 1, 2, *Frostbelt Automation: The ITI Report on Great Lakes Manufacturing* (Ann Arbor, Mich.: Industrial Technology Institute, September 1987).

48. Daniel Luria, Roland J. Cole, and Alan Baum, *When Industrial Policy Arrives: The Allocation of Manufacturing Extension* (Ann Arbor, Mich.: Industrial Technology Institute, 1992).

49. Irwin Feller, "American State Governments as Models for a National Science Policy," *Journal of Policy Analysis and Management*, Vol. 11, No. 2 (Summer 1992); Luria, Cole and Baum, *When Industrial Policy Arrives*.

50. Norsworthy and Jang, *Technology Adoption in U.S. Manufacturing*.

51. Michael S. Fogarty and Jar-Chi Lee, *Manufacturing Industry Clusters as a Logical Way to Structure Technology Deployment* (Cleveland, Ohio: Program Paper Series, The Center for Regional Economic Issues, Weatherhead School of Management, Case Western Reserve University, February 27, 1992).

52. NEMTC was established at Rensselaer Polytechnic Institute in 1989 as one of the first three Manufacturing Technology Centers under NIST's program. NEMTC is now a joint program of the New York State Science and Technology Foundation and Rensselaer.

53. Joshua Lerner, *The Small High-Technology Company, the Government, and the Marketplace: Evidence of the SBIR Program*, Discussion Paper 89-11 (Cambridge, Mass.: Science, Technology and Public Policy Program, John F. Kennedy School of Government, Harvard University, 1988).

54. Georgia Tech Research Institute, *Georgia Manufacturers Speak Out*; Susan Helper, *Supplier Relations at a Crossroads: Results of Survey Research in U.S. Automobile Industry*, Working Paper No. 89-26 (Boston, Mass.: Boston University School of Management); Industrial Technology Institute, *Frostbelt Automation*; NIST, *Overcoming the Competitive Barriers Facing Smaller Manufacturers*; Shapira, "Modern Times."

55. Kelley and Watkins, "The Defense Industrial Network"; Norsworthy and Jang, *Technology Adoption in U.S. Manufacturing.*

56. Committee on Technology Policy Options in a Global Economy, *Mastering a New Role: Shaping Technology Policy for National Economic Performance* (Washington, D.C.: National Academy of Engineering, May, 1993), p. 11.

57. NIST, *Manufacturing* (April 7, 1992).

58. Chapman, Clark, and Dobson, *Technology-Based Economic Development.*

59. Shapira, Roessner, and Barke, *Federal-State Collaboration in Industrial Modernization.*

60. ARPA, *Program Information Package for Defense Technology Conversion* (March 10, 1993).

61. NIST, *Manufacturing Technology Centers*; NIST, *The Manufacturing Technology Centers Program* (October, 1990); U.S. General Accounting Office (GAO) Report to the Ranking Minority Member, Committee on Small Business, U.S. Senate, *Technology Transfer: Federal Efforts to Enhance the Competitiveness of Small Manufacturers*, GAO/RCED-92-30 (Washington, D.C.: GAO, November 1991).

62. NIST, *Manufacturing Excellence Partnership*; ARPA, *Program Information Package for Defense Technology Conversion.*

63. National Institute of Standards and Technology, *A Catalog of U.S. Manufacturing Networks*, GCR 92-616 (Gaithersburg, Md.: NIST, September 1992); Gene R. Simons and M. Raghvachari, "Total Quality Control in Small Manufacturing Firms," *Transactions of the 45th Annual Quality Congress*, Milwaukee, Wisconsin, May 1991.

64. Shapira, Roessner, and Barke, *Federal-State Collaboration in Industrial Modernization.*

65. TECnet is a pilot electronic networking project for small metal-working firms that is funded by NEMTC, the State of Massachusetts, and NIST, and is located at Tufts University in Medford, Massachusetts. TECnet is under the direction of Dr. Leslie Schneider.

66. Jack Russell, *Challenges of Growth* (Airlie, Va.: NIST MTC National Staff Conference, March 31, 1993).

7

Research Universities and the Social Contract for Science

Harvey Brooks

Thus saying that new technologies have given rise to new sciences is at least as true as the other way around. And it is more on the mark to say that with the rise of modern science-based technologies, much of science and much of technology have become intertwined. This is the principal reason why, in the present era, technology is largely advanced through the work of men and women who have university training in science of engineering. This intertwining, rather than serendipity, is the principal reason why, in many fields, university research is an important contributor to technical advance, and universities as well as corporate labs are essential parts of the innovation system.

Richard R. Nelson, ed., *National Innovation Systems: A Comparative Analysis* (New York: Oxford University Press, 1993), p. 7.

For most of the period after World War II, universities, particularly the so-called research universities, constituted one of the center-pieces of U.S. science policy.[1] Although "universities proper" performed only 12 percent of national R&D (15 percent including separately organized and staffed FFRDCs that are administered by universities or university consortia), they performed more than 50 percent of all basic research, more than twice as much as any other performer, and they absorbed five times as much of federal obligations for basic research as industrial performers (including FFRDCs operated by industry), 52 percent of all such federal obligations.[2] A majority of doctoral scientists and engineers work in universities. The percentage of the nation's R&D performed in universities has increased fairly steadily, especially after 1960.

Within the last few years, however, there has grown up consider-able questioning of this *de facto* science policy, generally considered to have been launched in 1945 by the famous Vannevar Bush report[3] (and supported by the Steelman Report as well).[4] This questioning has included not only the role of universities *per se*,[5] but also more broadly the role of curiosity-driven and investigator-originated research in all R&D institutions, including the corporate laboratories of large multinational corporations.[6]

This chapter explores the role of universities in the nation's changing science and technology agenda and policy. Universities are, quite properly, struggling to protect their commitment to independent, investigator-originated scholarship while trying to make a more effective contribution to the public interest through collaboration with industry, where such collaboration is appropriate and builds on existing skills and strengths of the universities. While academic institutions as a whole, like all institutions that are governed collegially rather than hierarchically, are slow to change in response to top-down direction, their constituent parts are frequently quite opportunistic in responding to new opportunities. Today their difficulties are compounded by conflicting, sometimes unrealistic expectations as to what they can do to make the economy more competitive. At the same time, the political basis for balancing academic autonomy with these expectations is eroding.

Issues Regarding the Role of Universities

Among the issues that have been raised about the role of universities:

(1) Most important and new is the search for an enhanced role for academic research in contributing to the solution of the perceived problems of declining U.S. competitiveness in world markets. This is perhaps best symbolized by the creation of the National Science Board (NSB) Commission on the future of the National Science Foundation (NSF), intended to respond to this search.[7] NSF is faced with a choice. On the one hand, it could attempt to broaden its mission to take on the entire "technological food chain" from research to market. On the other, it could (as recommended by the NSB Commission) assume the lead in helping the Office of Science and Technology Policy (OSTP) to formulate a national science and technology policy involving all federal

agencies. In this policy NSF would assume a role which would be more appropriate and feasible in terms of its experience and external constituencies. At the same time it could become linked more effectively to the rest of the government science and technology (S&T) enterprise and ultimately to the entire national innovation system, including the private industrial sector.

From the Congress and elsewhere in government and industry the fundamental criticism of NSF, attributed to the predominant influence of its mainly university constituency, is that, by insisting on scientific autonomy and working primarily within established scientific disciplines on individual investigator-originated projects, academics are diverting scarce resources away from problems of greater importance to industry and those that lie at the interfaces between disciplines, which are often most relevant to the problems faced by society as a whole.[8]

(2) There is also a more general political disenchantment with universities. Congressional hearings on scientific fraud are symptoms of a loss of public confidence in the capacity of academic science to regulate itself.[9] There are also widely publicized allegations of lax accounting of indirect costs in universities or abuse of government regulations for indirect cost allowances.[10] Conflicts of interest of faculty who participate as principals in outside businesses and in consulting is another focus of criticism. Perhaps more subtle and profound is the loss of public confidence in what used to be seen as the traditional role of academic scientists as independent critics of the actions and policies of government and industry in their fields of expertise.[11] A perennial concern that predates the era of generous government support of university research is the alleged over-emphasis on faculty research and on prestige among disciplinary peers outside their own institution, to the detriment of teaching and the accessibility of faculty, especially senior faculty, to undergraduate students. Related to this is an accusation of a general neglect of values and ethics in favor of a kind of amoral rationality and cleverness as ends in themselves. This is often accompanied by allegations of a lack of attention to and respect for the values and culture of the Western intellectual tradition.[12]

(3) Congress has also criticized research universities for their collaborative relations with foreign-affiliated corporations and for

their encouragement of the growth in the number of foreign nationals participating in publicly supported academic research. These policies are alleged to result in the loss of "knowledge assets" paid for by U.S. taxpayers without fair compensation, to the ultimate detriment of the economic interests of U.S. citizens.[13]

(4) There is also increasing concern with the steady growth in the number of claimants for research support (including heavy involvement of foreign-born post-doctoral researchers and junior faculty), and the concomitant decline in the success rate of research proposals. This trend has occurred despite the fact that, on the whole, academic research funds have continued to grow in comparison with most other discretionary items in the federal budget.[14] There are also signs of growing resentment against scientists whose pleas for more academic research support sound more and more like the special pleading of a relatively well-off interest group.

(5) Another important concern has been the growth in the number and cost of "big science" projects and organized research centers supported by block grants, both of which are perceived to compete with investigator-originated and conducted project research. This has occurred as the number of attractive new scientific opportunities has outpaced the growth in available federal funding.[15] This concern has been further exacerbated by the rapidly growing practice in Congress of "earmarking" funds in appropriations bills for particular institutions and projects, usually large block-funded projects or facilities grants.[16] With tight budgets, these earmarked funds come directly at the expense of regular peer-reviewed programs in the affected agencies, and therefore also represent a direct attack on the autonomous decision-making mechanisms developed by the scientific community as part of the "social contract."

(6) Simultaneous with the rise in public and political skepticism about the value and conduct of academic research is an increase in skepticism by corporate managements about the value of corporate R&D not directly linked to product divisions and short-term business plans. There is increased tension within firms over the desire of technical staff (as well as some management) for the peer

recognition and good public relations that result from the scientific excellence of corporate R&D. This is seen as coming at the expense of the contribution of corporate R&D to the implementation of "bottom-line" business strategy—as a cost rather than a benefit to the business divisions that bring in the revenues that make corporate R&D possible. There has been a long-term decrease in general-purpose and long-term corporate research, while several famous corporate R&D laboratories, such as RCA, have nearly disappeared.[17] All these trends are reflections of the declining prestige of the kind of curiosity-driven, freewheeling research that was emphasized in the Bush report as a basis for a thriving economy and a healthy and peaceful society.

(7) Recently, with considerable prodding from the Congress, local politicians and laboratory administrators have begun to search for a new, more commercially-oriented role for large multipurpose national laboratories in the light of the declining importance of the national security mission which has accompanied the phasing down of the Cold War. This has resulted in competition between federal laboratories and universities to demonstrate their capacity to contribute to enhanced national economic competitiveness in order to forestall the threat of downsizing. (See Chapter 4.) As missions have declined, the laboratories themselves have tended to shift toward general-purpose research less distinguishable from that done in universities, but without the benefit to the economy, which universities provide, of a steady throughput of newly trained people into the private sector.[18]

(8) There have been major changes in the last fifteen years in the intellectual property regimes of both university and public laboratory R&D sponsored by the public sector. Linked with this has been a dramatic growth of jointly funded government-university-industry R&D centers (UIRCs) on university campuses. Partly paced by the emergence of biotechnology as a viable commercial industry closely linked to academic research, academia's tilt towards the industrial research culture has increased. This is often referred to in the science policy literature as the "privatization" of academic and public research.[19] This trend is also seen as potentially damaging to academic independence by many scientists,[20] and by some

politicians as a justification for forcing universities to extract social commitments from industry.[21]

Not all these phenomena or critiques are new to the late 1980s. There was a previous (and in some ways more intense) crisis of university research funding between 1967 and 1975, when academic R&D funding declined about 15 percent in real terms relative to its 1967 peak. The exception was biomedical research, which continued to grow, albeit at a reduced rate. The decline in academic R&D funding, especially in the physical sciences, was accompanied by a public demand, particularly among young people, for greater "relevance" of academic research to societal problems such as the environment and the Great Society—i.e., S&T problems as formulated in political terms by society, rather than defined in more technical terms by scientists. There was a growth of populist ("science for the people") and "public interest" science.[22] The brief decline was replaced in the second half of the 1970s by a new period of growth in both the numbers of R&D scientists and engineers that were employed and in federal support for academic research—a growth which continued until 1989. Simultaneously, there was a resurgence of growth in company-funded industrial in-house R&D.[23]

What can be inferred from this experience? One lesson seems to be that the fortunes of academic research and of pure science seem to be closely tied to overall societal demand for scientific and technological activity, as determined by factors largely external to science. This is contrary to the conventional wisdom of the academic science community, which tends to view applied technological activities, especially those funded by government, as in competition with the funding of basic research.[24] In the resurgence of the late 1970s, four factors were probably involved: the energy crises of the 1970s, which were accompanied by a remarkable intensification and diversification of energy R&D, the discovery of recombinant DNA techniques and the dramatic growth of new industrial applications of biology, the revival of the Cold War and the accompanying military build-up in the second half of the Carter administration, and the growth of computer science and engineering in universities with the maturing of the computer industry. The appearance of the nascent biotechnology industry on the scene was especially important because it provided a vivid demonstration of the unexpected utility of what had hitherto been viewed as the overly lavish support of "pure" research in molecular biology—a subject that industry had largely bypassed and that had been developed almost entirely in universities and public laboratories. It is not only that the public popularity of applied research benefited academic and basic research in its wake; it was also that the industrial and government development of technology were turning up entirely new intellectual issues, which became grist for the mill of basic research.[25]

The developments of the decade of the 1980s—especially the demand that university scientists accept more accountability for societal benefits from their research—have been widely interpreted both within and outside the scientific community as signs of the erosion and anticipated demise of the "social contract" between science and society, a social contract generally identified with the Bush report of 1945.[26] Such an interpretation seems questionable, however. It seems to be a consequence of some scientists clinging to a belief that the social contract is the only acceptable paradigm for the relationship of science to government, rather than a paradigm that can at best characterize only a small fraction of the total efforts of the scientific community. Thus the fundamental issue of public policy for science is not the validity of a single comprehensive social contract for complete autonomy in exchange for promised social benefits, but rather one of identifying what proportion of the overall national science and technology effort should be subject to the social contract in order to optimize its net contribution to society.[27]

Meanings of the Social Contract for Science and Society

The idea of a social contract between science and society has been predicated on an assumed rigid separation of science from politics. This idea goes back to the original Baconian program[28] and to the concept of science as a unique form of "public knowledge" introduced by John Ziman.[29] In Bacon's words: "nor can nature be commanded except by being obeyed and so those twin objects, human knowledge and human power, do really meet in one; and it is from ignorance of causes that operation fails." Similarly, Ziman argued that science is public knowledge in the sense that it is a social product generated collectively by a social system of open criticism, which is supposed to guarantee that the knowledge is asymptotically valid independent of the social context in which it arises.[30] The conviction that truth is independent of man's wishes and preferences easily extrapolates, in the ideology of the scientific community, to the belief that science (and even technology) should be generously funded but need not be controlled by society because it does not deal with values and power.

But this comes into conflict with other arguments used by the scientific community in justifying public support of basic science. These arguments stress the often unpredicted ultimate usefulness of unfettered curiosity-driven research. Yet scientists cannot have it both ways: ask to be subsidized by the public because of the ultimate practical value of their work, while arguing that science does not change power relations and, by itself, has no distributional effects.

This paradox can in principle be partly resolved by asserting that, while the results of science can be value-free, the agenda of science—the questions which it chooses to address—are not value-free, because they are inevitably influenced by the patrons of science as agents of society, and hence by politics and economics. While the questions that are asked determine the order in which discoveries are made and conclusions reached, and hence have political implications in the short run, it can be argued that these implications tend to be squeezed out as the structure of scientific knowledge is completed over time. Nevertheless, the more science is immediately useful to society, the more it is the present incomplete structure of science, rather than the ultimate finished edifice, that has present social consequences and thus cannot be considered wholly apolitical.

In the original discussions leading up to the Bush report, fear of the potentially dangerous influence of "politics" on the integrity of science was a pervasive theme. A typical expression of this fear is the following stricture of the Bowman Committee, the task force that Vannevar Bush charged with answering the third of the four questions addressed to him by President Roosevelt: "What can the Government do now and in the future to aid research activities by public and private organizations?" which included evaluating "the proper roles of public and of private research, and their interrelation." The Bowman Committee's apprehensions were formulated in the following stark terms:

> It is the firm conviction of the Committee that centralized control of research by any small group of persons would be disastrous; if this small group were backed by the power and prestige of the federal government and open to political influence, it would be catastrophic.[31]

It was this apprehension that led the Bowman Committee to recommend a federal institution "governed by a board of scientists and their sympathizers, who would choose—and control—a compliant federal director."[32] It was this recommendation that led ultimately to the Truman veto and a consequent five-year delay in the legislation that finally established the National Science Foundation (recommended by the Bush report).[33] Much the same argument was made more recently in the 1992 report of the National Academies of Science and Engineering, which recommended the creation of a Civilian Technology Corporation (CTC) to be funded with a one-time grant of $5 billion from Congress, but self-governed to minimize political influence. (See Chapter 3.)[34]

Much the same issue had been involved in the Polanyi-Bernal debate and the Marxist challenge to the claims of autonomy for science which had taken place in England during the 1930s.[35] Bernal had expressed the view that the present structure of science "is a structure of appalling inefficiency both as to its internal organization and as to the means of application to the problems of production and welfare."[36] There are parallels between this debate and the debate in the economic sphere between *laissez-faire* and economic planning, in the sense that the competition for recognition of priority of discovery in science was thought to provide a sort of "invisible hand" which semi-automatically maximized the rate of progress of science in the same way that the free market was supposed to maximize economic progress.

This debate continues to the present day. Robert Frosch is a former administrator of NASA who has spent most of his government career managing large government technology projects. This might have been expected to predispose him towards the planning perspective of Bernal. Nevertheless, he concludes that we are better off trying to make progress by being opportunistic in both science and technology. "Living with a rather diversified and complex apparatus," he says, "and being opportunistic inside it, is better than attempting to build formal policy frameworks which are likely to result in monsters that are even more difficult to live with than are the difficulties of a pluralistic system."[37]

A somewhat similar debate was launched in the 1960s by Alvin Weinberg, at that time director of the Oak Ridge National Laboratory, in a brave attempt to develop a better rationale for scientific choice, mainly in basic science.[38] This debate was reviewed at a conference near Paris in 1967, but not much progress towards resolution appears to have been made since that time.[39] The best conclusion may be some sort of healthy tension between planning and pluralism that never gets fully resolved.

Indeed, part of the problem of resolution arises because of the interpenetration of two rather different issues. One is the issue of autonomous planning by the technical community versus direction of science by forces external to that community. The other is the issue of centralized decision-making by a small number of "wise men" (as feared by the Bowman Committee) versus decentralized, opportunistic decision-making involving numerous players and the interplay of many forces, both political and scientific. These choices are essentially orthogonal to each other. Planning of science and technology can be either centralized or interactive and participatory but confined largely within the technical community, or it can involve external forces and actors but similarly be either centralized or bring in numerous representative social actors from both inside and outside the technical community.[40] Moreover, there is a virtual continuum along each dimension with various mixes of centralization and decentralization and various levels of representation of technical and non-technical people in decision-making bodies or participatory forums. In addition, there is the possibility of multi-stage decisions along each dimension, with broad strategies determined in a centralized manner and detailed tactics determined in a decentralized way, or with broad strategy determined with heavy participation of non-technical people and tactics determined largely by scientists. This is in fact the way the national biomedical research enterprise has evolved, which Frederickson refers to as a "culture warp" to denote its departure from the purity advocated by the Bowman Committee.[41] It seems almost certain that purity along either dimension is likely to be neither feasible nor desirable in the real world.

One of the pitfalls of the "social contract" is that what may have started out as a pragmatic condition for the efficient progress of science for the benefit of society—the largest accretion to conceptual knowledge for a given expenditure of effort and resources—may gradually evolve over time to be regarded as a "right" of the scientific community analogous to free speech or other human rights in democratic polities. In fact "freedom of research" is often spoken of in that light. The next step beyond freedom of research can be a notion that all scientists capable of doing good research are entitled to some sort of social support for their efforts. In an inaugural address as president of the American Association for the Advancement of Science, the Nobel laureate Leon Lederman seemed to adopt that viewpoint when he documented the growing number of capable scientists who were unable to pursue their research because federal agencies did not receive sufficient appropriations from Congress to support more than a small and declining percentage of "worthy" project grant applications.[42] This speech was immediately interpreted by many Congressmen, including many traditional friends of science such as George Brown, as a plea for a kind of scientific entitlement.[43] To an increasing degree in recent years, representatives of the scientific community, once regarded as impartial spokesmen for the public interest, have come to be viewed by politicians as just another "special interest group" in the same category as farmers, industry associations, or organized labor (which itself in the past used to be treated much more as a disinterested spokesman for the public interest and for greater equity in American society).

The rationale for the scientists' position was succinctly formulated in the 1960s by the sociologist of science, Bernard Barber: "However much pure science may eventually be applied to some other social purpose than the construction of conceptual schemes for their own sake, its autonomy in whatever run of time is required for this latter purpose is the essential condition of any long run applied effects it may have."[44] This comes close to being an article of faith in the academic science community, and becomes ingrained through the socialization of almost all scientists and many engineers trained in research at the PhD level. While there are many cases that seem to be consistent with the Barber thesis, supporting it with statistical evidence or hard analysis is another matter, and the rationale is still primarily a matter of faith.

However, it provides weak support for the social contract whenever there is a predisposition to challenge it in the external society on which scientists depend for support. Furthermore, the majority of scientists, and almost all engineers, work in environments where their agendas are largely determined from outside the world of science. They have, perforce, made their peace with working in a more utilitarian environment, where even the selection of basic research areas is mainly driven by some external perception of societal needs. Consequently they offer little sympathy to their academic colleagues who complain about the gradual erosion of their autonomy. As long as the relative global economic position of the United States continues to decline, and important technology-intensive areas of the public sector—notably the health care delivery system—continue to fall into disarray, the academic science community is in increasing danger of becoming politically isolated. Paradoxically, this isolation derives more from the success of science than its failures. In fact, the very contrast between the successes of basic science on its own terms, and the disarray in so many of the "downstream sectors" of the techno-economic system which use science, only serves to draw attention to the problem.

Other Challenges to Scientific Autonomy

There are also other more oblique challenges to scientific autonomy.

Growing Regulation and Bureaucracy

First, there has been a steady growth of legal and bureaucratic regulation of research in respect to the methods and techniques used in its conduct. This includes a host of regulations regarding the use of human subjects and laboratory animals in research, the application of industrial-style health and safety regulations to laboratory settings, regulations on the use of fetal tissue in research, and special rules relating to the use of recombinant DNA techniques. Many of these regulations are regarded as individually meritorious by researchers, but their sum total adds up to a level of bureaucratic oversight that is increasingly burdensome and is felt to slow research progress and to increase its cost (and hence to decrease the amount of research that can be supported under tight budgets).[45]

Social Regulation of Outcomes as Well as Methods

Much more controversial is the regulation of research in relation to its outcomes and conclusions. Some things are regarded as too dangerous to know because of the perceived likelihood of misuse by society or because they are thought to threaten public order or to undermine basic tenets regarding human nature that form the intellectual underpinnings of liberal-democratic societies. Examples of "forbidden knowledge" include the role of heredity in human behavior and social performance (the IQ debate, sociobiology, the identification of genes that predispose to criminality, even the role of genetics in predisposing to schizophrenia); genetic engineering; new reproductive technologies such as in vitro fertilization; and surrogate motherhood. The scientific community fears this sort of intervention as a "slippery slope" that opens the way to the thorough politicization of science in the future.[46]

Challenge to the Concept of Objectivity

Recently, there have been even deeper challenges to the very concept of "objective knowledge." It has become fashionable in some intellectual quarters to represent science as a pure social construction, reflecting the power relations and prejudices of the surrounding culture and existing political and economic structures.[47] An implication is that if the findings of science depend on power relations in the surrounding society, science should be consciously manipulated to modify existing power relations in favor of certain social groups, especially those that are seen as having been disadvantaged in the dominant Eurocentric culture. This is legitimate, it is argued, because science is already being manipulated by existing elites in support of existing power relations and material advantages.

Scientific Misconduct

A fourth challenge is the political issue of "scientific misconduct." It has recently triggered a debate both inside and outside the scientific community over the permissible limits of error in the dissemination of public knowledge. Several federal agencies have sponsored special offices to monitor record-keeping in scientific projects and to hear complaints of "whistle-blowers" in sponsored projects of the agency. An example is the Office of Scientific Integrity in NIH.[48] There is a great deal of debate within the scientific community as to the extent and nature of the problem and whether it is growing.[49]

Public and political sensitivity to the possibility of error or deliberate manipulation or fraud in the reporting of scientific work, and to the standards of proof required for general acceptance of results, is greatly increased when scientific findings have social consequences in regulation and public policy outside the scientific community itself.[50] This sometimes results in a tendency to legislate scientific procedures and models (for example, the most appropriate model of carcinogenesis) because society cannot tolerate the informality of the procedures of normal science in these areas where the application of knowledge in regulation and political decision-making has important distributional consequences.[51]

Procedural versus Substantive Accountability

Another growing issue is the increased procedural accountability often introduced into research projects as a surrogate for outcome accountability, since administrative and accounting procedures are more familiar to the lay administrators and Congressional committees that oversee science agencies. However, it also true that, because scientists claim exemption in the name of the social contract for certain kinds of procedural accountability, they may become more vulnerable to accountability for society's use of scientific outcomes over which they as scientists have little or no control.[52]

Two Parallel Competing and Cooperating Technical Cultures

Dasgupta and David distinguish between two different but interacting scientific cultures, to which they attach the labels "science" and "technology."[53] These are distinguished not by their content or cognitive skill requirements, but by the "socioeconomic rule structures under which the research takes place, and, most importantly, what the researchers do with their findings," as well as the audience to which they are primarily addressed. (To avoid confusion here I capitalize Science and Technology when they are used in a limited sense as labels for these distinct "cultures" of different research communities, rather than in their common meaning, recognizing that many individuals can act at different times or in different functions according to the norms of either culture.) In a somewhat idealized or caricatured sense, "the community of Science is concerned with additions to the stock of public knowledge, whereas the community of Technology is concerned with adding to the stream of rents that may be derived from possession of [rights to use] private knowledge."[54] Although industrial scientists and engineers are generally considered to be more concerned with "tacit knowledge" and academic scientists and engineers with "codified" knowledge, Dasgupta and David treat this difference, when it occurs, as "a derived consequence of the rules and reward structures that obtain in the different research communities" rather than as more fundamental differences between the Science and Technology communities.

The main advantage of the codified knowledge produced by the Science culture is that its costs of transmission and storage are much lower than for the knowledge produced in the Technology culture. Thus the "market" for Science information is much larger, as is the value, economic or non-economic, to individual users, but in the absence of "special contrivances such as intellectual property rights protection," the suppliers of Science information have no way of directly capturing economic benefit, either for themselves or to continue the support of their work. Instead, the reward system of the Science culture uses the coinage of reputation, awarded for priority of discovery and publication.

Of course, in the long run, cumulative scientific reputation is translatable into various economic benefits, such as higher salaries, academic promotion, and a greater variety and number of employment options. However, these economic benefits, though potentially large, are indirect and cannot be compared to the economic rents derivable from specific proprietary knowledge. Conversely, priority of conception is an important factor considered in the award of rights to proprietary knowledge with respect to human artifacts (though not natural phenomena, a distinction from the priority recognition in the Science culture, which includes natural phenomena). The reward system in the Science culture has an all-or-nothing character in the sense that the addition to reputation derives mostly from certified absolute priority. Although confirmation of a finding made by others may add something, it is generally minor. As Dasgupta and David point out in a rather sophisticated analysis, this characteristic of the reward system of the Science culture has certain sub-optimizing consequences for the allocation of effort and resources within that culture. On the other hand, the Technology reward structure often leads to much duplication of effort and high costs of information transfer that would have been unnecessary had the new knowledge originated within the Science culture.

Of course, the above description is a gross oversimplification, and in practice the costs and difficulties of transfer of both codified and tacit knowledge among Scientists, between scientists and members of other communities, including Technologists, and among Technologists are highly diverse, and have not been much investigated. Eric von Hippel has shown that Technologists in industry frequently exchange tacit knowledge that could ordinarily be considered as proprietary across company boundaries on an informal *quid pro quo* basis without objection from their superiors because mutual benefits are deemed to exceed competitive losses.[55] Furthermore, large companies find it economically beneficial to conduct a certain fraction of their technical activity according to the norms of the Science culture.[56] If this were not so, corporate R&D would make little contribution to the open scientific literature. There are examples of what appears to be a gradual convergence in practice between the Science and Technology cultures in academia and industry. In the last twenty years, however, there has also been a gradual shift of the center of gravity towards the Technology culture, especially evident in some fields in academia. This is also evidenced by a decline in corporate R&D in favor of R&D in product divisions in some of the modern high-tech industries, and by greater emphasis on acquisition of new knowledge externally through various sorts of alliances and patent exchanges in preference to in-house R&D.[57]

With respect to academia, for example, Cohen, Florida and Goe have documented the rapid growth during the 1980s of University-Industry Research Centers (UIRCs) jointly funded by federal and state governments, industry, and the universities themselves.[58] The aggregate budgets of more than 1000 UIRCs in more than 200 colleges and universities identified in their survey were estimated to exceed $4.5 billion, with about 30 percent of this funding from industry.[59] Of the UIRCs existing in 1990, nearly 58 percent had been founded in the decade of the 1980s, and 85 percent since 1960. About 16,000 graduate students and 9,000 undergraduates have some experience in these centers, and as many as 13 percent of all PhDs produced each year are likely to have had some association with such centers. This seems to represent a real revolution in the level of exposure of students and faculty to the Technology culture. On the other hand, it appears that the general culture of the UIRCs is still academic or Science. The surveys "suggest that UIRCs see themselves as principally providing windows on new technological developments" rather than "providing more direct contributions to...technical advance within industry." Nevertheless, 26 percent of UIRCs surveyed reported the "transfer of technology to industry" and "improving industry's products and processes" as important goals guiding their activities, and only 25 percent of UIRCs regarded these goals as unimportant. Only 16 percent of their R&D activity was reported as development, but basic and applied research each accounted for more than

40 percent of their R&D activity (in about equal percentages). Sixty-four percent of UIRCs reported that industry influenced their work moderately (36 percent) or strongly (28.5 percent).[60] These very recent data seem to indicate that some of the criticisms regarding the isolation of academia from industry, summarized in the introduction to this chapter, are out of date.

The survey also reveals some disturbing aspects of UIRCs, suggesting an erosion of the academic culture within them. For example, participating firms have the right to delete information from center reports in 35 percent of the centers, the right to delay publication in 52.7 percent, and the right to do both in 31.4 percent. There is no information on how often these rights are exercised or how much freedom of publication is actually affected, however.[61] Nor is there much data on the openness of these centers to all students and faculty on a non-discriminatory basis.

Dasgupta and David also express concerns about the "fragility" of the Science reward system relative to the Technology one, and fear that present trends may be an early warning of the danger of "predation of science by technology" even within academia. They assert that "the argument that if there is some useful research to be done it will be done within technology, and that it will be done there by cheaper means and without recourse to the public purse, betrays a staggering lack of understanding of the economics of science and technology." This seems an extreme view, but it might be a fair warning if political pressures were to force the National Science Foundation, for example, into taking on the entire "technological food chain"—a fear that was raised in the academic community by the formation of the National Science Board Commission, although in its report the commission itself warned against this. In a historical account of the development of university research in the United States, Rosenberg and Nelson have shown that American universities before World War II made important direct contributions to solving problems of local industries while at the same time institutionalizing and systematizing research in the "sciences of the artificial," especially in electrical and chemical engineering, but also in aeronautical, mining, and other branches of engineering which took on a life of their own, distinct from both industrial research and the traditional scientific disciplines.[62] Much of this development occurred in parallel with the rise of American universities to a world standard in several fields of pure research such as astronomy and nuclear physics in the second and third decades of the twentieth century, long prior to the great surge of American science after World War II. Thus, in this case, the two developments appear to have been complementary rather than competitive, even though there was little interaction between them.

Although Shapley and Roy have bemoaned the loss of prestige of engineers and engineering in the post-World War II era relative to pure science,[63] Rosenberg and Nelson suggest that they have considerably overstated their case, noting that "whatever its standing in terms of prestige, engineering is receiving more resources than the physical sciences...and research at medical schools is receiving far more resources than research conducted in the biology departments of Arts and Sciences schools." Furthermore, relative R&D resources devoted to engineering have steadily increased in the last two decades, as has the number of PhDs in engineering compared to science. The problem has been more that, unlike the rest of university science, academic engineering has been very much more dependent on military-oriented funding sources (about 60 percent of several of the largest disciplines), with the result that industrial connections of faculty and students have been biased towards the parts of industry that are not working primarily in the commercial sector.[64]

The growth of UIRCs in the 1980s seems to suggest that this situation is changing rapidly, and provides an institutional base for rapid adaptation of academic engineering to declines in defense spending, including R&D spending. The Clinton-Gore technology agenda includes significant emphasis on manufacturing research and education, which should assist both engineering and technical schools to support the shift to commercial innovation.

Indeed the worry expressed by Dasgupta and David, and echoed by a number of other economists who deal with the economics of innovation, is that "the growing dependence of economic growth upon the science-technology nexus has made the stability of this growth at high levels a hostage to what one would judge are possibly fragile features of the cultural and political environment." The danger is that universities will be diverted away from what they do well—namely the institutionalization and codification of the generic knowledge underlying industrial practice—to undertake more commercially oriented activities that require market judgments, which are not likely to be done well away from the business environment. Much the same concern attaches to the proposed commitment of the DOE weapons laboratories to industrial partnerships with commercial objectives, despite the fact that they have greater capability than universities in many "downstream" activities, including some specialized forms of manufacturing.

An essential point is that it is not only pure research that is to be identified with the Science culture. As explained by Rosenberg and Nelson, "the notion that nowadays academic research is mostly about things far removed from practical concerns, while widespread, is, quite simply, mistaken." In a recent report they say that:

Basic research involves the quest for fundamental understanding and, in the traditional natural sciences, such a quest has often been identified with research that was significantly distanced from immediate concerns with practical applications. However, a widely accepted definition of basic research has come to focus on the *absence* of a concern with practical applications, rather than the search for fundamental understanding of natural phenomena. This is indeed unfortunate, if not bizarre.

Indeed, much research in engineering and clinical medicine as well as in many other applied fields can be just as much concerned with fundamental understanding as the so-called "pure" sciences. This understanding may have more to do with man-made artifacts, tools, or procedures than with phenomena of undisturbed nature, but they are no less fundamental for that reason.[65] Furthermore, an increasing amount of such research may be concerned with the assessment of technology and its human and environmental impacts, which often requires deeper fundamental understanding than the original invention of technology.[66] The initial invention of technology can be and still is often carried out by trial and error with prototypes. (Even here, simulation and modeling, which require more fundamental understanding, are frequently more cost-effective, at least until the very last stages of development and deployment, when feedback from the market and the manufacturing floor become important factors in design.)

Even when fundamental understanding is not the principal objective, the Science culture may provide a more useful approach to technological problem-solving than the Technology culture, by insuring that the results are more "generic" and hence more easily communicated to a wide range of potential users. (See discussion of infrastructural science and technology in Chapter 3.)

Indeed, it has been suggested in the past that the development of generic "technological capabilities" for their own sake with only the most general idea of their ultimate application might be a worthwhile goal for national science policy in addition to basic research. This might be called "pure technology" as a kind of complement to "pure science" and would often arise from technological opportunities perceived as a result of discoveries in pure science. Search for applications of high-temperature superconductivity might be an example. In most cases, development of such generalized technological capabilities may most efficiently be undertaken within the framework of the Science culture rather than the Technology culture.[67]

The identification of basic research with lack of applicability or usefulness or with motivation by pure curiosity is a gross oversimplification that has done a considerable disservice to the implementation of science policy in the political sphere. The idea that the applications of science cannot be foreseen is often taken much more literally by lay people than by scientists. Indeed, the entire notion of government support for pathbreaking technologies (see Chapter 3) is actually predicated on the notion that potential applications can be foreseen in general terms as they emerge from science.

In fact, there are very few fields of research where the choice of the questions to be asked and the tactics to be followed in pursuing them can be determined exclusively by the conceptual structure of the subject. In a few subjects like particle physics or cosmology or some areas of pure mathematics, the strategy of research can probably be defined only by considerations internal to science. But in the more usual case of fields like condensed matter physics, most of chemistry, and most of the environmental sciences, research opportunities are much more numerous and diverse than can be chosen by internal considerations alone. Some additional notion of potential applicability, no matter how vague and tentative, is necessary to construct a viable research strategy. Furthermore, in some of the very fields which seem most "pure," considerations of cost minimization and the design of an efficient measurement technology are essential in plotting research strategy. Although the expected results may be of purely intellectual interest, the means that have to be devised to arrive at these results may have important and unforeseen fallout in other areas of science or technology, even though one could not justify the cost of the research in terms of the fallout alone.

If the actual fallout applications could have been foreseen, they could have been pursued much more cost-effectively by working on them directly, but more frequently such potential fallout does not become apparent until well along in the project, and it still requires considerable additional investment to realize the potential. Such challenging projects in pure science also often develop human skills among the participants that can be later applied in other more "practical" areas.[68] The pure science projects serve as a screening device for identifying people with unusual skills and creativity, some of whom go on to become participants in important technological developments.[69]

Although basic research, and increasingly much applied research, and even some development, involve the creation and dissemination of public rather than appropriable knowledge, there are costs of communication of this knowledge in a form in which it can be understood and appreciated by potential users. The conventional treatment of basic research as a pure public good is thus a gross oversimplification. The transfer of knowledge from such research requires a reasonable match in level of technical sophistication between the supplier and the receiver. This sophistication depends on tacit knowledge that is not codified in readily communicable form, even for basic research, and generally speaking, this tacit knowledge can be derived only from actual participation of the recipient in R&D in a reasonably closely related field, including personal interactions within a research network or "invisible college." That this is really so is suggested by research on citations to basic research papers in patents pertaining to important industrial innovations; the citations tend to come predominantly from universities in geographic proximity to the company where the invention originated, and overwhelmingly from the same country.[70]

It also follows from the above considerations that in areas where potential industrial applications are of interest it is not enough for academic research just to be generally relevant to the industry. Both the industrial and the academic researchers must be part of a single community with numerous linkages and exchanges, even when they are operating in different "cultures" in the sense of Dasgupta and David.[71]

Recommendations for the Future

(1) There is a need to continue building on the progress that has already been made in cultivating closer information exchange between academia and industry, but this does *not* mean that academics should attempt to imitate industrial research or development. The primary goal of academic research should still largely be the understanding or the development of generic design methodologies and tools rather the development of products for a market. However, this understanding can and should include understanding of industrial technology beyond short-term and product-specific problem-solving. It is important to preserve the essentials of the Science culture of public knowledge even when working on problems of industrial or commercial interest in the academic setting. Any attempt to solve problems which are application-specific or product-specific in the academic setting should be viewed as a way of illuminating or testing broader principles and methods of design or invention. Information exchange between industry and academia can and should be facilitated by how the knowledge emerging from the Science culture is packaged, and an important goal should be the further improvement of the effectiveness of this exchange using the tools of modern information technology as well as improvement of the tools themselves.

(2) There should be a larger role for academia in "infrastructural" research and technology (in the sense used in *Beyond Spinoff*, and also see Chapter 3),[72] that is, knowledge-seeking driven by problems arising out of technological development rather than arising just from the conceptual structure of science *per se*. This could also

serve as a stimulus of new areas of basic research resulting from reformulation of problems uncovered in technological development or technology assessment, which could be efficiently attacked in the style and with the reward structure of academic research. The research would thus seek deeper understanding than required by the immediate needs of the original industrial problem, ideally giving rise to new questions of conceptual importance which might not otherwise have been discerned. The further generalization and packaging for optimal and low-cost dissemination should also be one of the objectives of technology-driven academic research. All federal agencies should be encouraged to fund such research in universities according to their expertise and industrial contacts. This applies particularly to National Institute for Standards and Technology (NIST), which should develop an extramural program in universities (much as the National Institutes of Health [NIH] have parallel but separate intramural and extramural programs). In fact, in the long run the Advanced Technology Program (ATP) of NIST should probably be reoriented to put greater emphasis on infrastructural research, in which universities might play a larger role than at present.

(3) The function of universities in the national research system should remain mainly the creation of "public knowledge." The relatively modest level of "privatization" of university research that has already taken place in the last decade should not be encouraged to grow. Furthermore, there is a need to develop a consensus among government agencies and universities on appropriate policies and practices for industry-sponsored research in universities to avoid erosion of the Science culture. In particular, the conduct of proprietary research for particular companies and restrictions on freedom of publication and open access to research results should be closely circumscribed.

Better industry-university interaction might be achieved without erosion of the academic culture through the development of "buffer institutions," that is, separately organized and administered research centers with their own staff, career lines, and reward systems rather than industry-funded centers that are integral parts of the university. University faculty, research staff, and students could have access to these separate centers and be employed by them on a part-time basis, but would remain answerable to the university in their career development, in contrast with the core staffs of the research center. These "buffer institutions" would be something like affiliated teaching hospitals or the government-sponsored FFRDCs now supported mainly by the Department of Defense and the Department of Energy. The recommended separation would impede erosion of the academic culture through preserving the reward system of universities while permitting further growth in communication between the industrial and academic cultures.[73] The financial incentive for the university would be that such centers could provide a source of partial support for students and other university personnel, as well as access to specialized equipment on a subsidized basis. In the long run it seems likely that some of the federal support now going to mission-oriented FFRDCs should be shifted to such university-affiliated buffer institutions, since the economic competitiveness mission inherently has less need for large-scale laboratories and also needs wider dispersion than the present federal missions. Such institutions are not without their problems, however, and considerable experimentation would therefore be required to find the best arrangements.

(4) There is a need for many more working scientists and engineers in universities to learn more about what solid scholarship has to tell us about the innovation process. It is particularly important for engineers, biomedical researchers, and other applied researchers in universities to collaborate with economists and management experts in studying the innovation process itself as a challenging research problem. A prerequisite for this may be intensification of foundation and government support for interdisciplinary scholarship in this area, accompanied by an effort to develop literature and teaching materials directed particularly to the education of bench scientists and engineers in the results of this scholarship. Not only could this lead to more sophisticated rationales for the support of academic science, but it could also stimulate more academic interest in how basic research results could be packaged for communication to technologists and policy-makers.

This would also increase the number of people in the universities who were qualified to design and participate in the industrial extension services being expanded under the new policies (and discussed at length in Chapter 6). The academic scientific community cannot afford to sit back and become hostage to the dysfunctional performance of parts of the overall science-utilization system over which it believes it has no influence. An influential segment of the community of working scientists needs to become more sophisticated, informed, and influential participants in the national dialogue about U.S. economic performance and not simply leave it to the economists and management experts.

Concluding Remarks

Rosenberg and Nelson conclude their report on the role of universities with the following sentences:

A shift in emphasis of university research toward more extensive connections with the needs of civilian industry can benefit the universities if it is done in the right way. That way, in our view, is to respect the division of labor between universities and industry that has grown up with the development of the engineering disciplines and applied sciences, rather than one that attempts to draw universities deeply into a world in which decisions need to be made with respect to commercial criteria. There is no reason to believe that universities will function well in such an environment, and good reason to believe that such as an environment will do damage to the legitimate function of universities. On the other hand, binding university research closer to industry, while respecting the condition that research be "basic" in the sense of aiming for understanding rather than short term practical payoff, can be of enduring benefit to both.[74]

I am in general agreement with this statement with two important provisos: First, the shift should occur by providing new opportunities to faculty in universities to come forward with new projects of their own devising that satisfy broadly defined criteria designed with the advice of industry advisors. The peer review of individual projects should be conducted with strong, though not exclusive, participation of appropriate industry scientists and engineers. Second, the shift in the overall research portfolio should be gradual enough to permit orderly adjustment within a mix of projects that still leaves considerable scope for research driven primarily by scientific opportunity arising from the conceptual structure of science, and not just from well-defined or well-characterized societal or industrial needs.

Notes

1. Harvey Brooks, *The Research University: Centerpiece of Science Policy?* Working Paper WP 86-120 (Columbus, Oh.: The Ohio State University College of Business, December 1986).

2. National Science Foundation, *1992 Science and Technology Pocket Data Book*, NSF 92-331 (Washington, D.C.: Division of Science Resources Studies, NSF, 1992).

3. Vannevar Bush, *Science—The Endless Frontier*, A Report to the President on a Program for Postwar Scientific Research (Washington, D.C.: Office of Scientific Research and Development, July 1945; repr. National Science Foundation, 1980), pp. 5–40.

4. John R. Steelman, Chairman, *Science and Public Policy: A Report to the President*, Vol. 1: *A Program for the Nation* (Washington, D.C.: U.S. GPO, August 27, 1947).

5. For examples of recent reports, see: *Renewing the Promise: Research-Intensive Universities and the Nation* (Washington, D.C.: A Report Prepared by the President's Council of Advisors on Science and Technology [PCAST], U.S. GPO, December, 1992); *In the National Interest: The Federal Government and Research-Intensive Universities* (Washington, D.C.: A Report to the Federal Coordinating Council for Science, Engineering, and Technology [FCCSET] from the Ad Hoc Working Group on Research-Intensive Universities and the Federal Government, U.S. GPO, December 1992); Government-University-Industry Research Roundtable, *The Future of the Academic Research Enterprise: Report of a Conference*, James D. Ebert, Chair (Washington, D.C.: National Academy Press, 1992); Erich Bloch, *Fateful Choices: The Future of the U.S. Academic Research Enterprise: A Discussion Paper* (Washington, D.C.: Government-University-Industry-Research Roundtable, National Academy Press, March 1992).

6. John Simon, "The R&D Dilemma: The High Cost of Cutting Back," *The Harvard Business School Bulletin*, April 1993, pp. 34–37 (Report of a Harvard Business School Conference on the Future of Industrial Research, organized by Professor Richard S. Rosenbloom, February 1993).

7. National Science Board Commission on the Future of the National Science Foundation, *A Foundation for the 21st Century: A Progressive Framework for the National Science Foundation*, NSB-92-196, November 20, 1992 (William H. Danforth and Robert Galvin, co-chairs).

8. For a very early articulation of such a viewpoint in a somewhat different political context, see A.M. Weinberg, "But is the Teacher also a Citizen?" *Science*, Vol. 149 (1965), pp. 601–606; for a more extended critique along these lines, see Daniel Alpert, "Performance and Paralysis: The Organizational Context of the

American Research University," *Journal of Higher Education*, Vol. 56, No. 3 (May/June 1985) pp. 241–281.

9. David H. Guston, *The Demise of the Social Contract for Science: Misconduct in Science and the Nonmodern World*, Working Paper No. 19 (Cambridge, Mass.: Program on Science, Technology and Society, Massachusetts Institute of Technology, 1992).

10. Report on Activity of the Committee on Energy and Commerce, House Report 102-1096, Part 2, December 31, 1992; see also Oversight and Investigations Subcommittee, Rep. John D. Dingell, chair, "Hearing on Indirect Costs," March 1991.

11. Jerome R. Ravetz, *Scientific Knowledge and Its Social Problems* (Oxford, U.K.: Clarendon Press, 1971; Penguin Books, 1973).

12. Allan Bloom, *The Closing of the American Mind* (New York: Simon and Schuster, 1987).

13. Congress, "Is Science for Sale? Conflicts of Interest vs. the Public Interest," Hearings before the Human Resources and Intergovernmental Relations Subcommittee, Committee on Government Operations, House of Representatives, 101st Cong., 1st Sess. (Washington, D.C.: U.S. GPO, June 13, 1989); see also E.B. Skolnikoff, Chair, MIT Faculty Study Group appointed by the Provost on the International Relations of MIT, "The International Relations of MIT in a Technologically Complex World" (Cambridge, Mass.: MIT, 1992).

14. Daryl E. Chubin, project director, *Federally Funded Research: Decisions for a Decade*, OTA-SET-490 (Washington D.C.: U.S. Congress, Office of Technology Assessment, May 1991).

15. Christopher C. Hill, "Considerations in Funding Large-Scale Science," *CRS Review*, Vol. 9, No. 2 (February 1988), pp. 3–5; Elizabeth Baldwin and Christopher T. Hill, "The Budget Process and Large-Scale Science Funding," *CRS Review*, Vol. 9, No. 2 (February 1988), pp. 13–16.

16. See, for example, Thomas Toch and Ted Slavsky, "The Scientific Pork Barrel," *U.S. News and World Report*, March 1, 1993, p. 58.

17. Richard Florida and Martin Kenney, *The Breakthrough Illusion: Corporate America's Failure to Move from Innovation to Mass Production* (New York: Basic Books, 1990); David A. Hounshell, "Industrial R&D in the United States: An Exploratory History," paper prepared for conference on the future of industrial research in the United States, February 10–12, 1993.

18. Federal intramural laboratories absorb nearly three times as much federal obligations for basic research as does industry. See NSF, *1992 Data Book*, p. 4.

19. Yaron Ezrahi, *The Descent of Icarus* (Cambridge, Mass.: Harvard University Press, 1990); see Part III, "The Privatization of American Science," pp. 237–290; Partha Dasgupta and Paul A. David, *Towards a New Economics of Science*, CEPR Publication No. 320 (Stanford, Calif.: Center for Economic Policy Research, Stanford University, 1992).

20. Letter from Ernest M. Henley, President, American Physical Society, to NSB Commission on the Future of NSF, NSBC-393, October 14, 1992.

21. As examples of Congressional pressure on NIH to have its grantees require limits on the pricing of drugs produced by firms having licenses from universities based on patents obtained as a result of NIH grant support: "Extension of remarks" by Ron Wyden, *Congressional Record*, Thursday, March 11, 1993, 103rd Cong., 1st Sess., Vol. 139, No. 30, 139 Cong. Rec. E612; "Research Deals Eyed," *Newsday*, February 9, 1993, p. 16; President's Health Task Force Hearing, Pharmaceutical Panel (Washington, D.C.: George Washington University, March 29, 1993).

22. Harvey Brooks, "Can Science Survive in the Modern Age?" *Science*, Vol. 174 (October 1, 1971), pp. 21–30; Harvey Brooks, "Physics and the Polity," *Science*, Vol. 160 (April 26, 1968), pp. 396–400; Don K. Price, "Science at a Policy Crossroads," *Technology Review*, Vol. 73, No. 6 (April, 1971), pp. 3–9; Don K. Price, "Purists and Politicians," *Science*, Vol. 163 (January 3, 1969), pp. 25–31.

23. See Harvey Brooks, "National Science Policy and Technological Innovation," pp. 119–167, in Ralph Landau and Nathan Rosenberg, eds., *The Positive Sum Strategy: Harnessing Technology for Economic Growth* (Washington, D.C.: National Academy Press, 1986).

24. I predicted this development in the Donald Hamilton memorial lecture given at Princeton in 1973. See Harvey Brooks, "Are Scientists Obsolete?" *Science*, Vol. 186 (November 1974), pp. 501–508 .

25. See Harvey Brooks, "Lessons of History: Successive Challenges to Science Policy," pp. 11–22; and Brooks, "The Future: Steady State or New Challenges?" pp. 163–172, in Susan E. Cozzens, Peter Healy, Arie Rip and John Ziman, eds., *The Research System in Transition*, NATO Advanced Study Institute, Series D, Vol. 57 (Dordrecht, Netherlands: Kluwer Academic Publishers, 1990).

26. Henry Etzkowitz, "The National Science Foundation and United States Industrial and Science Policy," to appear in the British library journal *Science and Technology Policy*, February 1993; Guston, *The Demise of the Social Contract for Science*; Deborah Shapley and Rustum Roy, *Lost at the Frontier: U.S. Science and Technology Policy Adrift* (Philadelphia: ISI Press, 1985).

27. Harvey Brooks, "Knowledge and Action: The Dilemma of Science Policy in the 70s," *Daedalus: The Search for Knowledge*, Vol. 102, No. 2 (Spring, 1973), pp. 125–145.

28. The term "technological program" has been used to describe Bacon's philosophy by Donald A. Schon, *Technology and Change* (New York: Delacorte Press, 1967), see esp. chap. 8. This is further discussed in, for example, Harvey Brooks, "Can Science Survive in the Modern Age?" *Science*, Vol. 174 (October 1, 1971), pp. 21–30.

29. John M. Ziman, *Public Knowledge: An Essay Concerning the Social Dimensions of Science* (Cambridge, U.K.: Cambridge University Press, 1968).

30. Hobbes observes how geometry (and, by implication, other sciences) "cross no man's ambition, passion or lust"; quoted in Sheldon S. Wolin's introduction

to Marlan Blisset, *Politics in Science* (Boston: Little Brown and Company, 1972), p. ix of Foreword.

31. Quoted in Donald S. Frederickson, "Biomedical Science and the Culture Warp," December, 1992, to be published in W.N. Kelley, M. Osterweis, and E. Rubin, eds., *Emerging Policies for Biomedical Research* (Washington, D.C.: Association of Academic Health Centers, 1993 Health Policy Annual III) . The original reference is the Bowman Committee report on federal aid to research activities, 1944–45, Office of Scientific Research and Development files.

32. Quoted from Frederickson, ibid.

33. J. Merton England, *A Patron of Pure Science* (Washington, D.C.: National Science Foundation, 1982), esp. chaps. 2–4.

34. COSEPUP Panel on the Government Role in Civilian Technology, NAS/NAE/IOM, Harold Brown, Chairman, *The Government Role in Civilian Technology: Building a New Alliance* (Washington, D.C.: National Academy Press, 1992).

35. Stephen Turner, "Science as a Polity," presented at the annual meeting of the Society for Social Studies of Science (4S), Minneapolis, Minn., 1990. I am indebted to David Guston for calling my attention to this reference, a draft of which he provided. This paper gives an excellent account of the history and contemporary intellectual context of the Bernal-Polanyi debate; see also J.D. Bernal, *The Social Function of Science* (London: Routledge, 1939); Michael Polanyi, "The Republic of Science," *Minerva*, Vol. 1, No. 1 (1962), pp. 54–72.

36. Bernal, ibid., p. xiii, as quoted in Partha Dasgupta and Paul A. David, *Towards a New Economics of Science*, CEPR Publication No. 320 (Stanford, Calif.: Center for Economic Policy Research, Stanford University, October 1992).

37. Robert A. Frosch, "Relevance, Irrelevance and General Confusion: Problems in Science Policy," 15th J. Seward Johnson Lecture in Marine Policy (Woods Hole, Mass.: Woods Hole Oceanographic Institution, January 3, 1983).

38. For a summary of Alvin Weinberg's contributions to this debate, see A.M. Weinberg, *Reflections on Big Science* (Cambridge, Mass.: MIT Press, 1967); A.M. Weinberg, "Science, Choice and Human Values," *Bulletin of the Atomic Scientists*, Vol. 22 (1966), pp. 8–13. For a general review and critique building on this debate, see Harvey Brooks, "Models of Science Planning," *Public Administration Review*, Vol. 31, No. 3 (May–June 1971), pp. 364–374.

39. Harvey Brooks, "Can Science Be Planned?" reprinted from *Problems of Science Policy: Seminar at Jouy-en-Josas on Science Policy* (Paris: OECD, 1967).

40. For a proposal to develop this kind of participatory planning, see Carnegie Commission on Science, Technology, and Government, *Enabling the Future: Linking Science and Technology to Societal Goals*, September 1992.

41. Frederickson, "Biomedical Science and the Culture Warp."

42. Leon Lederman, "Science, the End of the Frontier?" (Washington, D.C.: American Association for the Advancement of Science, 1991).

43. George E. Brown, Jr., "A Perspective on the Federal Role in Science and Technology," pp. 23–34, in Margaret O. Meredith, Stephen D. Nelson, and Albert Teich, eds., *AAAS Science and Technology Yearbook, 1991* (Washington, D.C.: American Association for the Advancement of Science, 1991).

44. Barber, *Science and the Social Order* (New York: Collier Books, 1962), p. 139.

45. Task Force on Science Policy, *The Regulatory Environment for Science*, Science Policy Study Background Report No. 10 prepared for the Committee on Science and Technology, House of Representatives, 99th Cong., 2nd Sess. (Washington, D.C.: Office of Technology Assessment, U.S. GPO, 1986).

46. Robert S. Morison and Gerald Holton, eds., *Limits of Scientific Inquiry*, *Daedalus*, Vol. 107, No. 2 (Spring 1978); see esp. Loren R. Graham, "Concerns about Science and Attempts to Regulate Inquiry," pp. 1–12.

47. For example, Bruno Latour, *Science in Action* (Cambridge, Mass.: Harvard University Press, 1987); for a recent political flirtation with this view, see George E. Brown, Jr., "The Objectivity Crisis," *American Journal of Physics*, Vol. 60, No. 9 (September 1992); for a comprehensive critique of these intellectual trends, see Gerald S. Holton, "The Value of Science at the 'End of the Modern Era'," prepared for presentation at the 1993 Sigma Xi Forum, *Ethics, Values, and the Promise of Science*, February 26, 1993.

48. Guston, *The Demise of the Social Contract*.

49. National Academy of Sciences, *Responsible Science: Ensuring the Integrity of the Research Process*, Vol. 1 (Washington, D.C.: National Academy Press, 1992).

50. Harvey Brooks and Chester L. Cooper, *Science for Public Policy* (New York: Pergamon Press, 1987). See especially p. 3; and also Brian Wynne, "Uncertainty—Technical and Social," Chap. 8, pp. 95–113.

51. Brooks and Cooper, ibid., pp. 108–11; Sheila Jasanoff, *Risk Management and Political Culture: Social Research Perspectives*, Occasional Paper on Current Topics No. 12 (New York: Russell Sage Foundation, 1986).

52. David Guston, private communication.

53. Dasgupta and David, *Towards a New Economics of Science*, pp. 15–20.

54. Ibid., p. 16.

55. Eric von Hippel, "Cooperation between Rivals: Informal Know-How Trading," *Research Policy*, Vol. 16 (1987), pp. 291–302; Stephan Schrader, "Informal Technology Transfer between Firms," *Research Policy*, Vol. 20 (1991), pp. 153–170.

56. Nathan Rosenberg, "Why Do Companies Do Basic Research (With their Own Money)?" unpublished manuscript, Stanford University, January, 1988; see also R.R. Nelson, "What is 'Commercial' and What is 'Public' about Technology, and What Should Be?" in Nathan Rosenberg, Ralph Landau, and David Mowery, eds., *Technology and the Wealth of Nations* (Stanford, Calif.: Stanford University Press, 1992), pp. 57–71.

57. Simon, "The R&D Dilemma."

58. Wesley Cohen, Richard Florida, and W. Richard Goe, *University-Industry Research Centers in the United States: A Report to the Ford Foundation* (Pittsburgh: Center for Economic Development, Carnegie Mellon University, June 1992).

59. Sources of support were as follows: industry, 30.7 percent; university, 17.7 percent; state government, 12.3 percent; federal government, 34.1 percent; other, 5.2 percent.

60. Cohen, Florida, and Goe, *UIRCs in the U.S.*

61. Ibid., Table 19.

62. Nathan Rosenberg and Richard R. Nelson, *American Universities and Technical Advance in Industry*, Center for Economic Policy Research Discussion Paper Series No. 342 (Stanford, Calif.; Stanford University, May 1993).

63. Shapley and Roy, *Lost at the Frontier.*

64. Alic, et al., *Beyond Spinoff*, Chapter 4.

65. Rosenberg and Nelson, *American Universities*, p. 21. For a further discussion of the basic-applied distinction based on research strategy, see Harvey Brooks, "Basic and Applied Research," in *Categories of Scientific Research*, NSF 80-28 (Washington, D.C.: papers presented at a National Science Foundation Seminar, National Science Foundation, 1980).

66. Harvey Brooks, testimony printed in *National Science Policy*, pp. 136–154, Hearings Before the Subcommittee on Science and Astronautics, U.S. House of Representatives, 91st Cong., 2nd Sess., No. 23 (Washington, D.C.: U.S. GPO, 1970). See p. 145: "I would like to emphasize my belief that our national capability for technology assessment and for the wise management and direction of technology to improve the quality of life is as critically dependent on a strong and broadly advancing basic science as was our military posture in the fifties and early sixties."

67. Baker, et al., "Organization of the Government for Science and Technology," Report of the Ad Hoc Committee of the National Science Board," May 2, 1962, revised January 1963; see especially: Appendix, "Steps Toward the Answers." The proposal here comes quite close to the concept of "pathbreaking technology" as explained in *Beyond Spinoff*, pp. 384–390, although the latter concept is not necessarily restricted to technological ideas suggested by basic research into natural phenomena. See also Chapter 3 of this book.

68. Harvey Brooks, "Returns on Federal Investments: The Physical Sciences," paper commissioned for the workshop on *The Federal Role in Research and Development*, November 21–22, 1985, NAS/NAE/IOM, NAS Academy Industry Program; Keith Pavitt, "Why British Basic Research Matters (to Britain)" January 18, 1993, draft for paper based on presentation in the ESCR Seminar Series: *Innovation Agenda*, made at the Institution of Civil Engineers, London, December 3, 1992; R.R. Nelson and R. Levin, et al., "The Influence of University Science Research and Technical Societies on Industrial R&D and Technical Advance," Policy Discussion Paper Series No. 3 (New Haven: Yale University, 1986); Levin,

et al., *Yale Survey on Appropriability and Technological Opportunity* (New Haven: Yale University, 1991).

69. Keith Pavitt, "What Makes Basic Research Economically Useful?" *Research Policy*, Vol. 20, pp. 109–119. According to Pavitt, many people trained to advanced levels in pure science "go on to work in applied activities and take with them not just the knowledge resulting from their research, but also the skills, methods, and a web of professional contacts that will help them tackle the technological problems that they later face."

70. Rebecca Henderson (MIT), Adam Jaffe, and Manuel Trajtenberg, "Telling Trails out of School: Basicness, Appropriability and Agglomeration Effects in the Generation of New Knowledge: A Comparison of University and Corporate Research with the aid of Patent Data," March 1991, draft only. Similar conclusions have been reached by Mansfield and by Pavitt. See, for example, Keith Pavitt, "What Makes Basic Research Economically Useful?"; Edwin Mansfield, "How Do We Measure What We Get When We `Buy' Research?" *The Scientist*, August 19, 1991.

71. Richard R. Nelson, editor, *National Innovation Systems: A Comparative Analysis* (New York, Oxford: Oxford University Press, forthcoming 1993), especially Chapter 16.

72. Alic, et al., *Beyond Spinoff*, Chapter 12, pp. 390–393.

73. Harvey Brooks, 1970 House testimony printed in *National Science Policy*.

74. Rosenberg and Nelson, *American Universities*, pp. 44–45.

8

Putting People First: Education, Jobs, and Economic Competitiveness

Dorothy S. Zinberg

What good is it to log on if you can't read?
Congressman George Brown, chair, House Committee on Science, Space, and Technology, in a speech to the AAAS Colloquium, Washington, D.C., April 16, 1993.

No one can select from the bottom those who will be leaders at the top, because unmeasured and unknown factors enter into scientific, or any, leadership. There are brains and character, strength and health, happiness and spiritual vitality, interest and motivation, and no one knows what else, that must needs enter into this . . . calculus.
Vannevar Bush, *Science—The Endless Frontier*, A Report to the President on a Program for Postwar Scientific Research, p. 24.

On February 22, 1993, President Clinton and Vice President Gore announced their new program, *Technology for America's Economic Growth: A New Direction to Build Economic Strength.*[1] The goals for their technology policy include:

•Long-term economic growth that creates jobs and protects the environment;

•A government that is more productive and more responsive to the needs of its citizens; and

•World leadership in basic science, mathematics, and engineering.

This chapter addresses the ways in which some of these goals—particularly those related to education and jobs—might be achieved, and urges that the goal of "world leadership in basic science, mathematics, and engineering," as outlined in the Clinton-Gore program, be recast as the more modest goal of "approaching" rather than "achieving" world leadership across the board in all fields of science and mathematics. If the goal is not reached as promised, public cynicism about the efficacy of government proposals will deepen. As the data make shockingly apparent, the current state of education in the United States, particularly in grades from kindergarten through high school (K-12), and especially when compared with the accomplishments of other industrialized countries and industrializing countries such as China, makes "world leadership" in all of these fields an unachievable near-term objective and one that even in the long term is probably not desirable in a global economy.

This chapter assesses the Clinton-Gore objectives by examining the current state of education, the quality of the workforce, and the skills and intellectual qualifications needed to create and to fill the new technology-based jobs in the decades ahead. It also examines the perceived mismatch between jobs and skills, and between the expectations of the workforce and the realities that the country faces. As is becoming increasingly apparent in the ferocious international battle for technology's products and markets, the contributions made by human capital and intellectual resources are crucial to the economic vitality of the country.

Secretary of Labor Robert B. Reich and Secretary of Education Richard W. Riley have written widely about the tight coupling needed between education and the economy in order to achieve the nation's goals for global competitiveness. With President Clinton and Vice President Gore, they have been pushing forcefully for new options for financing college education and for creating apprenticeship (and national service) programs that would adapt the long-established models employed with success in Germany and Sweden.

This chapter emphasizes the social revolution in values and attitudes required to bring about the much-needed changes in the workforce, changes that entail a redefinition of the value of work itself. To date the initiatives regarding programs that will strengthen the school-to-work transition are most frequently described as "for those students who do not go to college." The "non-college" label assigns second-class status to the would-be apprentices—those left over from the elites who attend college—exactly the opposite of what was intended by the framers of these breakthrough proposals.

To establish the idea that a technical track is equally valuable to the individual, ideas about employment, compensation, and the needs of the country must undergo a radical revision of opinions, attitudes, and expectations, as a would-be meritocracy re-examines the need not only for college graduates but also for a differently trained, technically skilled workforce which, since World War II, has lost status and income to a college-educated populace. Erasing from the national vocabulary the negatively value-laden term "vocational training" will mark the first step toward redefining the route to a promising career. For example, the state of Oregon, among others, has introduced "professional-technical training" as the title of its new state-wide high school programs.[2] Obviously a new name quickly becomes a euphemism for business-as-usual, unless it is accompanied by a meaningful restructuring of existing programs which must include innovations not only in coursework but in teaching and career-guidance counseling as well. Together the name and demonstration programs can provide the new metaphors and symbols for the first steps needed to bring about attitudinal changes toward once-devalued "vocational" training.

If there is an American dream, a large part of it is fulfilled by a college degree, or by working to be able to send one's children to college. If the rules of the game are changed, and traditional college education is to be redefined as valuable for some young people but not others, there must be a social-cultural agreement that assigns equivalent status and income to those who choose technical training. Meanwhile, the training must be redesigned as professional training to fulfill aspirations as well as adequate preparation for jobs. Obviously, part of the solution will lie in a better link-up between apprenticeships and higher education (and industry), a move already underway in many community colleges. However, the community college system, despite its impressive successes since World War II, still is not seen as fulfilling the dream of the four-year college. In order to revitalize technical training, the community college, as the data demonstrate, must become integral to the apprenticeship program and a demonstrable stepping stone to a four-year degree. Empowering community colleges to assume additional vital roles in education (including life-long learning) will bring about changes that will extend to traditional colleges and research universities. All of these institutions will be competing for students, tuition fees, and funding from government and private sources. The results are likely to transform the very nature of higher education itself.

In a recent Arkansas study of a proposed apprenticeship program that would begin in high school and could lead to an associate of arts degree, parents and students voiced strong opinions about the need for the credits to be transferable to a four-year degree-granting college, because they believe a college degree is the "minimum requirement for the job market."[3] Arkansas ranks seventeenth in the country in high school graduation rates, but below the median for individual earnings. If feelings about the value of a college degree are so pronounced in Arkansas, they are undoubtedly reflected throughout most of the country.

Consequently, the apprenticeship and national service programs that would abet the president's technology policy statement should take into consideration the profound symbolic significance of a college education and the need to attach an equal significance to alternative routes to achievement.

However the new workforce is to be educated and trained, its critical place in achieving economic goals is by now well established. *Fortune,* a magazine long dedicated to the hows of amassing said "fortune," recently celebrated thinking on its cover. The banner heading, *Brain Power: How Intellectual Capital Is Becoming America's Most Valuable Asset,* declared that brain power "can be [America's] sharpest competitive weapon."[4] Not surprisingly, these sentiments are echoed by the head of the United States Steel Workers:

U.S. workers need to be viewed as part of the answer to U.S. competitiveness problems, not part of the problem. . . . [Too many people] think of them as a liability. . . . People thought technology would somehow solve our problems and we did not need to worry about people. Modern technology is most effective in the hands of skilled workers.[5]

Workers' skills must be available in what Lester Thurow has described as the "hot" industries of the twenty-first century—microelectronics, biotechnology, the new materials industries, telecommunications, civilian aviation, robots and computers. He points out that "these industries could be anywhere. Where they actually will be depends on who organizes the top-to-bottom brain power to capture them. In the twenty-first century, the winners will be those who understand this new reality."[6]

The New Security: Economic Competitiveness

Since World War II, U.S. leaders have measured national security in numbers—the numbers of nuclear warheads, the numbers of troops deployed in Europe, and, most important, the numbers attached to the defense budget.

By the mid-1980s, the public, ahead of the experts in anticipating social change, began to interpret open-ended questions about national security as questions about economic security. By 1988, almost 60 percent of Americans surveyed indicated that "our economic competitors like Japan are greater threats to our national security than our military adversaries like the Soviets."[7] By the time the former Soviet Union unraveled, concerns about national security had shifted to anxiety about declining U.S. economic competitiveness. Only 23 percent of a national sample of Americans believed that the United States was still the key economic power in the world; and a startling 90 percent had some degree of "serious" concern about the country's ability to compete effectively in the world economy.[8]

Admiral William J. Crowe, Jr. (ret.), chairman of the Joint Chiefs of Staff under Presidents Reagan and Bush, gave a powerful endorsement to the shift in public opinion. He wrote to the *New York Times* in 1992: "There is no doubt in my mind that our national security depends, first, on our domestic strength—a strong economy, a stable industrial base, social unity, and educational excellence."[9]

Education as the Linchpin of Economic Security

The National Competitiveness Act to be introduced in 1993 reinforces the commitment of Admiral Crowe and of the Clinton administration to education and economic competitiveness. The bill authorizes key federal agencies such as the Departments of Commerce, Defense, and Energy, and the National Science Foundation to increase federal support of K-12 science education and to initiate technical apprenticeship programs in order to dramatically improve the quality of the workforce. Education and economic planning are firmly embarked on a new but unsteady partnership as arguments over the deficit and the budget threaten to derail many of the proposed changes.

As Japan's dazzling economic victories appeared to be tied significantly to its high literacy rate and its well-trained, disciplined workforce, the search was on for the key to their success and the reasons why the United States lost both its lead in the very technologies it had invented, and its once secure markets.

Japan's economic productivity, as well as that of South Korea and Taiwan, appeared to be built on the high educational achievements of its students in computer sciences, engineering, mathematics, and other technical subjects. And Germany, whose export trade leads the world, has a dual-track school system: a three-year apprentice program for non-college students combines on-the-job training in a factory and theoretical education in school. The status of these students is high—markedly better than that of students tarred with the "vocational" label in the United States. The result is a workforce marked by efficiency and meticulous standards commensurate with the high wages and esteem with which they are rewarded.

The full impact of a well-trained workforce became apparent as U.S. market shares plummeted. Obviously, there were other culprits, such as the weak links between product development and manufacture, or outdated management practices. Nevertheless, an ill-educated and inefficient workforce that could not keep pace with technology induced a major reexamination of the nation's schools. Belatedly, the vital coupling between education and the economy became painfully apparent and new efforts to integrate economic competitiveness with education were, if not initiated, at least explored by virtually every institution and organization in the country. Whether in the Department of Defense, the National Science Foundation, private foundations, new commissions, universities, state and local governments, PTAs, industry, or individuals with ideas, education has emerged as the new criterion for national economic security.

Many American leaders and educators had been blasting sirens for several decades, warning the country that its once vaunted system of education, particularly in grades K-12 and in technical workers' training, was in serious trouble. But for many years they were preaching to each other. After all, the university system was (and still is) the best in the world, and Americans continue to garner a lion's share of Nobel prizes. Less apparent was the population that came to be known as "the forgotten half," the economically disadvantaged youth of the country. They were not being well served in their K-12 schooling, and the all-important links between schooling and employment, as demonstrated in the training programs in Germany, Sweden, Japan and other industrialized countries, were notably missing. The American workforce, which had become uncompetitive with Third World workers because of higher pay, also lost out to industrialized countries with even higher pay scales.

Most of the early warnings went largely unheeded until the economy began to sag dramatically as market shares were lost in electronics, automobile manufacture, machine tools, and dozens of other products once the domain of American enterprise.

Among those who recognized the symptoms of decay was Governor Jim Hunt of North Carolina, who in 1984 encouraged the Carnegie Foundation to launch a major effort to explore the interdependence of economics and education.[10] Three influential works emerged within a decade: *Workforce 2000* in 1987; *The Forgotten Half* in 1988; and *America's Choice: High Skills or Low Wages!* in 1990.[11] Together, they helped invigorate reading and mathematics programs and school-to-work transitions at local, state and national levels. The new theme, as succinctly laid out by the Commission on Youth and America's Future (1991), states: "America cannot afford to waste one student."[12]

By 1989, President George Bush joined the governors under the chairmanship of then-Governor Bill Clinton of Arkansas to issue a promise that "By the year 2000, U.S. students will be first in the world in science and mathematics achievement." Now, in 1993, two members of President Clinton's cabinet, Secretaries Riley and

Reich, have put forward legislation incorporating the national education goals that had been promulgated in the earlier Bush-Clinton initiative, legislation that recognizes the need to link schooling with skills that young people will need in order to be fruitfully employed and to restore the prosperity of the country.[13]

A recent survey of 500 American leaders in the public, private and nonprofit sectors revealed that education was the number one concern for both Democrats and Republicans, while defense, which a decade ago dominated the national agenda, dropped to a low of twentieth. Relations with the former Soviet Union as a subject of concern fared only marginally better, ranking nineteenth.[14]

At a recent meeting of the nation's leading economists, the only issue on which they were able to suppress their "bickerings" was the need to "educate the workforce" as a means by which human capital would be strengthened and the nation's economic growth rate spurred.[15]

Beyond that agreement in principle, the consensus would undoubtedly crack. There are sharp differences of opinion and conflicting policy proposals—the relative roles of local, state, and the federal government must be reconsidered—but the new "consensus" on the indispensable role education must assume provides the energy and drive for the debate over the many routes that will be proposed.

Education, now joined by training, has moved quickly to the top of the list of determining variables in the new battle for economic competitiveness. The interdependence of national security, economic competitiveness, and educational achievement has become apparent to a large enough segment of government, industry, and education (although not to enough of the public) that the need to move simultaneously on all fronts is clear.

The United States has a history of taking action when goaded by external forces. After the Soviets launched Sputnik in 1957 there was a pause to reexamine educational policy. The result was a massive infusion of funds to train scientists and engineers. Or, as the dean of the Harvard Graduate School of Education described it:

The focus was on academic excellence for talented and privileged children, with not much attention to the average and below-average child. Then in the '60s there was a big push toward access and equity for the entire population, with a focus on disadvantaged children. Today's reform movement, in lots of ways, is trying to combine both by establishing rigorous standards not just for the talented kids but all kids. It is being driven in part by the business community that is making the argument that we simply need to train all children at a much higher level so that we will be in a position to compete internationally. In a sense, then, the focus on excellence is back, but with a much broader mandate than existed in the late 1950s.[16]

Human resources have become the bedrock of the economy in industrialized countries that are increasingly knowledge-based rather than resource-based societies. Information technologies have indeed led to automated machines, but they require scientists and engineers to invent and improve them, managers who know how to put them to work, and a workforce that has the competence to work with "thinking machines." In short, everyone along the line has to measure up.

Human resources have not always been considered the bedrock of an economy. But as work has shifted from brawn to brains in many economically critical endeavors, the vastness of the Soviet Union and the insignificant landholdings of Japan and the "Little Tigers" Hong Kong, Singapore, Taiwan, and South Korea, have demonstrated that the size of land holdings is no longer the defining variable for economic strength. Many formerly affluent African economies have suffered as once uniquely-valuable natural resources such as cotton or chromium are increasingly being replaced or displaced by new technologies and synthetics.

New technologies, particularly the information technologies of fax, e-mail, and computers, along with now seemingly old-fashioned tape recorders that helped pave the way for Khomeini's dramatic return to Iran, or the copying machine that abetted the overthrow of communist regimes in Eastern Europe, have made national boundaries permeable to knowledge and capital. Transportation technologies have advanced, allowing people to move around the world economically and quickly. Human capital now moves across national borders like much of the technology that has quickened the pace of the flow of knowledge. As a result, there is a world market not only for products, technology, and capital, but for labor as well.[17] These changes have brought about radical shifts in social, economic, military, and political arrangements, marked by an unprecedented degree of international interdependence.

Through foreign investments and electronic transfer of funds and information, industries and corporations once embedded in their own nation's economy have created a global economy along with a global workforce. U.S. businesses now have more than $1.2 trillion in assets abroad; by the end of the 1980s more than a third of U.S. multinationals' earnings came from overseas operations. According to Robert Reich, "'American' corporations and 'American' industries are ceasing to exist in any form that can meaningfully be distinguished from the rest of the global economy.....The standard of living of Americans, as well as the citizens of other nations, is coming to depend less on the success of the nation's core corporations and industries...than it is on the *worldwide demand for their skills and insight* [emphasis added]."[18] Consequently, matching the education and skills of the workforce (and government policy) with the needs of industry is a leading requirement for national productivity in a global economy.

Focusing on the Technical Workforce

As the Clinton-Gore policy recognizes, implementing technology entails enlisting the brain power not only of the elites in science, engineering, technology and management, but also of the technical workforce. Harvey Brooks wrote:

I am inclined to feel that, by and large, the elite will take care of themselves. The problem we face in the future concerns much of the rest of the population, the millions who will have to operate and supervise the technologies on whose reliable, safe and predictable performance all of us will increasingly depend...the airline pilots and mechanics, the tanker captains, the safety inspectors, the construction engineers, the medical technicians, the design and manufacturing engineers, and the shopfloor workers.[19]

To this list should be added their teachers and those whom they supervise.

Skilled workers carry out demanding operations. Once, their jobs were defined by physical exertion; now the demands are likely to be characterized by the mental exertion of mastering technical literature and symbol manipulation.[20] Not long ago shopfloor workers could learn how to operate equipment by watching an experienced worker. Now machines operated by computer chips or similar electronic devices require the operator to be able to comprehend hundreds of pages of complicated computer manuals, and, in addition, to be able to advance intellectually with the rapid changes in technology. If technical students are not trained to learn from manuals, the apprenticeship programs proposed by the Clinton administration will be more protracted and expensive than those in countries such as Japan, South Korea, China, and India, where the technical workforce has greater skills in mathematics and computer sciences and will continue to outperform their American counterparts.

Of course, American engineers are developing new technologies in corporate and government laboratories which they then transfer to engineers in a product development or production division. But to a large extent the competitiveness of the American firm will be a function of its technologically sophisticated workers. Without radical improvements in their education and training, they could prove to be the weakest link even in an otherwise efficient system. A skilled workforce generates jobs, as well as a rise in living standards, in national income, and in global influence. Its skills "essentially become a firm's only long-run strategic source of competitive advantage."[21]

The distance between the cognitive skills acquired by technically trained students and what corporations require seems infinite if it is examined by looking at the current educational situation in American schools, and even more distressing when viewed in its international context.

For more than three decades, the interest and achievements of American students in science and mathematics, and more recently in quantitative subjects in general, have been dropping at an alarming rate. In 1991, the Department of Education reported that only 5 percent of students graduating from American high schools know enough mathematics to take advanced college mathematics courses or to handle the technological jobs on which the nation's economy increasingly depends. In addition, the "achievements" of high school graduates do not reflect the sad reality that 25 percent of high school-aged youth have already dropped out of school. The 1992 U.S. Census reports that the percentage of all teen-age students who dropped out of high school or were placed in classes at least one year below their age group had risen from 29.1 percent to 34 percent between 1980 and 1990. Were dropouts included in the final figures on mathematics achievement for their age group, the percentage of those performing behind their age group would be significantly higher.[22] In the borough of the Bronx in New York City, the only high school that has an on-time graduation record of more than 50 percent is the Bronx High School of Science. Overall in New York City only 38.6 percent of students get their diplomas within four years. Only 57 percent of the students graduate by the time they are 21 years old, the last year they can remain in high school.[23]

In one inner city school, for example, "only 17 percent [of the students] are in a college-preparatory program. Twenty percent are in the general curriculum, while a stunning 63 percent are in vocational classes [that lack up-to-date electronic equipment], which most often rules out college education." Of these, only 25 percent will graduate.[24] On a national basis the figures are equally dismaying: In 1990, 35 percent of teen-age students (up from 30 percent in 1980) had dropped out of high school or had been placed in classes at least one year below their age group.[25]

A report from the National Center for Manufacturing Sciences adds to the depressing statistics:

Today's young Americans spend barely 9 percent of their first eighteen years in school. . . . Less than forty percent of the nation's high school students use computers at school. (Japan's Ministry of Education plans to equip all Japanese schools with computers by 1994. Nearly all [98.5 percent] of Japan's high schools already have computers.). . . Estimates are that by 1993, 93 percent of the largest U.S. firms will be teaching employees the three R's and other basic work skills. . . . A prominent education expert estimates that only 7 percent of U.S. high school seniors—are receiving a high-quality education.[26]

The international comparisons are even worse. When American twenty-three-year-olds were rated for their performance in mathematics and science against Koreans, Canadians, and Europeans of the same age, Americans finished last on every test.[27]

Nine- and thirteen-year-old American students currently rank 15th in science and 16th in mathematics internationally, behind Canada, China, England, France, Germany, Italy, Japan, Russia, South Korea, and Taiwan—only Portugal and Jordan score lower. Assuming that the data for the Department of Education study were collected in 1990, that means that roughly one-fifth of those who will graduate by the year 2000 are already in school—in a seriously flawed system. And one-third of the mathematics and science teachers in American schools are not qualified to teach those subjects.

Among the general population, 60 million people read at a level "less than equal to the full survival needs of the society." That is one-third of the entire adult population.[28] Other estimates raise the figures closer to 85 million people who are either "functionally incompetent" or who "just get by."[29]

The statistics on American educational deficiencies have taken on repetitious tones. When the numbers are examined in a political and social context, they reveal some of the profound problems with which the country must grapple. Clearly it is not only the "forgotten half" but the more fortunate half of American students as well who have fallen behind world standards.

The effects of even the most dramatic K-12 educational reforms on the quality of the workforce will, of course, be much too slow in appearing to make any short-term contribution to national competitive performance. An inadequate education system will degrade economic performance long before a sufficient number of well-educated school leavers entering the labor force can positively affect its productivity and performance. Thus the retraining of the existing labor force, along with a concomitant reorganization of work to take full advantage of people as they gain upgraded skills, should receive a higher priority. The Clinton-Gore program does include a massive commitment to manufacturing extension. (See Chapter 6.) The problem is that the program may be too focused on firms with "hard" technology, and not enough on realizing the full potential of technology through exploitation of both individual and collective worker skills and entirely new forms of organization involvement and "social learning" on the job.[30]

But First, Where Are the Jobs?

This chapter might well have been titled, "Yes, but . . . ," because for every observation there is often a confounding, even contradictory piece of evidence or a new development for both college-educated graduates and those with high school or some college background.

Predictions five years ago by the National Science Foundation of a dire shortage of 675,000 scientists and engineers by the year 2006 now appear inflated. Today the job market is difficult even for graduates of universities as prestigious as the California Institute of Technology, where master's and PhD degree graduates received 31 and 42 percent fewer job offers, respectively, than between 1990 and 1992.[31]

Despite this alarming trend, statements predicting major shortfalls of scientists and engineers have continued to appear: "A serious shortage of engineers is a distinct possibility by the year 2000, caused by falling numbers of engineering graduates (down by 9350, or 12 percent since 1986) and the retirement of the large cohort of engineers who entered the profession after World War II."[32] By 1991 mathematicians were using words like "disaster" and "catastrophe" to describe the glut of young PhDs on the job market. Hiring freezes and budget curtailments at state universities, combined with declining enrollments, turned predictions upside down: the University of California at Los Angeles (UCLA) received "a mind-boggling" 1800 applications for three or four temporary positions.[33] In addition, as many as 300 well-trained Russian mathematicians sought employment in the United States within the past two years alone, and the numbers continue to rise, while some 40,000 Chinese students educated in the United States show little inclination to return home.[34] On the other hand, the countervailing trend is reflected in the return home or to jobs outside the United States of more than 25,000 American-educated South Koreans and Taiwanese, on whom many universities and industries have become dependent.

The glut-versus-shortage debate has been raging for almost a decade. The report entitled *Workforce 2000: Work and Workers for the Twenty-First Century* has dominated policy discussions with its assumptions of a "labor shortage" or a "skills mismatch," and that the American economy would "grow at a relatively healthy" pace. The growth in jobs was predicted to be largely in the service economy, thereby demanding more high-skilled professional and technically trained workers. Because the workforce is deficient in people with the requisite training, the result would be a serious labor shortage, it argued.[35]

More recently, however, another report, *The Myth of the Coming Labor Shortage: Jobs, Skills, and Incomes of America's Workforce 2000*, has challenged the conventional "labor-shortage" wisdom.[36] This study argues that rather than producing more college graduates, "the bigger and more important challenge is to improve the jobs, pay, and skills of the non-college educated workforce."[37] What it has added to the equation is the need for integrated policy making. But even if education and training for higher skills were to be successful, without the appropriate jobs such training would prove of little value. This "supply push" must be combined with policies that simultaneously take into account the types of jobs being created by the economy.[38] One sensible formulation of the skills-versus-jobs debate states: "The problem is not a short supply of skills for the kinds of jobs that presently exist, but scarcity of skills required in the kinds of jobs *that will have to be created* if the nation's economy is to regain its competitive edge."[39] Until policy makers resolve the differences between forecasts of a glut versus a shortage of skilled workers and take heed of the need to create different jobs, policy formulations will have little chance of being translated into effective action.

Fundamental Changes in the Workforce

The forecast of a crippling shortage was assumed to be fact until radical changes that affected the workforce itself began to be dramatically expressed. A faltering economy and a sharp downturn in the defense industry have led to large-scale unemployment and dislocations not only in defense but in the banking, aerospace, and construction industries. In California alone, an additional 100,000 jobs are expected to be lost in the closing of military bases scheduled for 1993.[40] Thousands of engineers, technically skilled personnel and managers in these industries have been peremptorily dismissed. (The state university system, as in most states, has been equally bruised. In 1992, enrollment was cut almost eight percent when funding for higher education was reduced nine percent. More than 6500 classes have been discontinued, and 1500 faculty laid off, and enrollment has dropped by 22,000 in the past two years.)[41]

Defense employment dropped from 7.2 percent in 1970 to 5.1 percent of national employment of 118.4 million people in 1991. By 1992 it had dropped to 4.3 percent. If there are additional large, sustained cuts, some 2.5 million defense-related jobs could be gone by 2001.[42] The major loss in employment from downsizing the military is likely to be felt by minorities, for whom the educational opportunities in the military services have provided training in technical skills as well as college credits and degrees.

New terms such as "re-engineering" or "right-sizing" have burst into print. They mean, in effect, that an organization can be more productive with fewer people; it is the workers who are redundant, not, as predicted, the job opportunities. For the most part, workers will be unable to return to their original jobs regardless of an upswing in the economy. Only 15 percent of recently unemployed workers expect to return to their jobs, whereas 44 percent returned after the last four recessions.[43]

Automation, long anticipated to reduce the number of people needed to run an organization efficiently, has finally become a potent factor in the economy. It affects workers at many levels, whether they are skilled in nuts and bolts, management, marketing, distribution, or even with the very technologies that have led to their unemployment in the economy.[44]

Job bumping is another effect of re-engineering and the recession—once more with deleterious effects on those on the lowest rung of the education chain. Surplus PhDs along with master's and bachelor's degree-holders in engineering and technical subjects are dropping to jobs which in the past required education one or more levels below their formal training. According to a report in the *Economist*, approximately one-fourth of all bachelor's degree graduates in the United States are working in jobs that do not require a college degree.[45] What does that portend for non-college graduates, who comprise more than 50 percent of the population? Or for those whose education leaves them without any marketable skills? Undoubtedly, low-skilled workers with technical training from inferior high schools or those whose training has not been updated are even more vulnerable to unemployment than they were before the most recent recession.

In addition, the new, lean organizations favor the hiring of "contingent workers," who are temporary and part-time, and therefore not entitled to benefits or other forms of job security.

Traditionally, large-scale unemployment has been a malady of factory workers. But several years ago white collar jobs also began to disappear in large numbers. *The Wall Street Journal*, in a series of articles about jobs in the United States, "Down the Up Escalator," has reported that the absence of jobs is the issue—not an acute labor shortage. Both high- and low-paying jobs are now being transferred overseas.[46] Although it remains a matter of dispute among labor specialists as to how many and how significant the engineering and computer programming job losses are, it is clear that information-technology companies are contracting work to professionals in Russia, Ireland, Israel, India, Malaysia, Singapore, and China.

It is difficult to reconcile the economic forecasts that predict a labor shortage by the year 2000 with today's employment situation. As *The Wall Street Journal* observed sadly, "Millions of American workers are embarking on a journey of insecurity, and professional and highly skilled workers are not exempt."[47] John Gibbons, now the science advisor to President Clinton, wrote in 1991: "Our education and skilled workers are such that Mexican workers making a fourth of our wages or less are just as skilled and productive and error free as our own workers....Executive compensation, on the other hand, is twice the world average for other industrial countries."[48] (Perhaps executive compensation will be the next target for downsizing.)

Despite these gloomy statistics, other data demonstrate the need for more skilled workers. Are there simply different skills that will be needed? Or is the United States facing the dilemma of knowing that in the short run there will be a surplus of certain skills, but in the long run the country's economic competitiveness depends on workers with greater reasoning and mathematical skills who can master the complexities of the new process technologies? Getting there requires nothing less than a "social revolution in education and skills."[49]

Lester Thurow has written that the United States "needs a workforce where the top twenty percent have the skills to invent the new product but in addition, that another two-thirds can work on all the related processes of getting the product to market. This requires that more than 85 percent of the population be trained more at a very different level."[50]

To be able to maximize the potential brain power in the United States requires a framework—a shared system of values which confirms the significance of education and work for the individual and the nation. The Clinton-Gore goals require a national consensus consonant with the "new reality." This poses a daunting challenge to a nation which, despite its enormous strengths and accomplishments in science and technology, is struggling with a faltering economy and a debilitating national debt—and little consensus on the direction to take in resolving either.

The American Dream: A College Education

To develop a solid apprenticeship program will require that the American dream concerning education, status, and work is not forestalled. For more than fifty years, college education has been the ticket to upward mobility. The "new reality" reveals that the dream is not being realized for more than half of American society. Much of this population, particularly in the inner cities, feels left out, embittered, or even worse, hopeless and despairing. Yet these are the very people on whom will rest the country's success or failure to implement many of the new technologies needed to regain economic competitiveness. When Vice President Gore waxes lyrical about his favorite projects—supercomputers and "information superhighways"—the questions of who will be able to operate and service them, and who will be able to use the information they generate, should be given as much consideration as who will create, design, engineer, and manufacture them.

A technically trained workforce will command a crucial role in making these super-technologies work, but that is the very population that is ill-prepared to move into the jobs generated by them.

Not only is education for employment in the information technologies deficient, but high school students' interest in colleges of engineering, computer studies, or other technical programs is also diminishing. An annual study of career aspirations among Pittsburgh-area high school students (college and non-college bound) yielded some perplexing findings.[51] Of the top twenty-five choices, lawyers were first: the popular TV program "L.A. Law" was credited with putting them at the head of the list. Not one of the top twenty-five choices reflected an interest in science, engineering, or technology. Musician, fashion designer, model, actor, and hair stylist all ranked well above electrical and computer engineer. Hardly the stuff of integrated circuits and information superhighways!

Estimates vary about the benefits of a college education in lifetime earnings for an individual, but there is little doubt that the differences are significant. In 1987, white males between the ages of 26 and 65 who had 16 years of education earned somewhere between 40 to 50 percent more per year than those with 12 years of education.[52] In 1991, Senator William Bradley observed that "a college graduate will earn about 60 percent more than someone with a high school diploma."[53] More recently the Bureau of Labor Statistics reported that the average wage of a male with a bachelor's or higher degree was $51,804; with a high school diploma, $27,865. The comparable figures for women are $33,615 and $19,309.[54] (Regrettably, women still earn significantly less than their male peers for the same or equivalent work.) Even if the figures are inflated, one of the powerful determinants for choosing to go to college is self-evident.

Senator William Bradley also observed, "Our economy rewards college graduates because we need their skills so deeply."[55] As the United States comes to grips with its "new reality," the economy should prepare to reward technically skilled workers because we need *their* skills so deeply.

New programs for technical trainees must be geared to eradicating the caste distinctions between technical and college work without in any way undermining the ambitions and opportunities for those who aim to enter college. The training must become a positive option, instead of a dead end, as it is currently perceived. To give substance to the rhetoric about the value of vocational/technical training, it must lead to employment compensation that will be closer to that of the college graduate over a lifetime.

With the Clinton-Gore plans to create needed apprenticeships and national service programs, unless the administration can narrow the divide by making non-college training and careers as valuable in status and income as the careers of college graduates, it runs the risk of widening the gap between those with college educations and those who have been left behind.

Achieving social and psychological acceptance of the European-type apprentice programs—that is, programs tied to industrial training but with the American add-on of community college courses—will require not only the "social revolution in education," but an even more difficult social revolution in values and attitudes. The American dream of college has been so successful in providing a commonality of middle-class membership that any deviance from the goal risks a sense of failure, particularly for minorities and immigrants for whom a college degree has served so demonstrably as the passport out of poverty.

Any new emphasis on technical training threatens to withhold an implicit promise to recognize, nurture, and reward talent through formal higher education. The new workforce will have to lose the dream unless it can be provided with opportunities for advancement through community, junior, or eventually four year colleges. These integrated programs (the integration would have to include tightly-coupled links to industry) would contribute to diminishing the social distance between those for whom a high school diploma (or less) is a terminal degree, and the college graduate.

The new programs are likely to change the nature of traditional colleges, whose teachings have grown out of the humanities and the social and natural sciences. As all but the elite colleges and universities struggle to fill their classrooms, financially pressed educational institutions will begin to offer a broader range of technical courses, thereby becoming more like the community colleges with whom they will be in competition. Such a transformation would exert more pressure for "relevance" on already beleaguered departments. Literature and philosophy, non-commercializable sciences as well as the social sciences, will have to be protected in the search for relevance. Efforts toward the revolutionary social changes that would make possible the next steps in revitalizing the workforce will have to enlist social anthropologists, psychologists and sociologists, themselves threatened by unemployment. In short, the intellectual achievements of the more affluent days of higher education must also be nurtured and protected from market forces, in order to make the empowerment of technology part of a balanced educational system, not one that is forsaking one tradition for the sake of meeting a new national need.

In *The Rise of the Meritocracy*, an English sociologist wrote, "War woke people up to the fact that the nation possessed a supply of ability never ordinarily used to the full."[56] In the postwar United States that awareness led to the G.I. Bill, which has been described as "a massive investment in human capital, an investment of a size and a scope never before contemplated . . . the single most important element of the stunning post war recovery . . . the domestic counterpart of the Marshall Plan."[57] Not only did it provide a *de facto* industrial policy over two decades; it also increased the scientific and technical workforce as a percentage of the total labor force by several times more than any of the country's industrial competitors.[58] World War III or its functional equivalent, some historians have argued, is being waged now, only this time it is an economic war.[59]

The G.I. Bill also gave reality to the American dream of a college education. This has become the symbol of achievement, of social mobility of people between and within vaguely defined socio-economic groups. If the "melting pot" did not live up to its promise, a college education did. It provided scientists and engineers with an opportunity to learn something about the humanities and the social sciences and a greater average lifetime earning capacity. Today if college is not within reach, the future appears foreclosed and faith in the country diminished.

President Clinton's national service program has been devised to keep that option as well as the technical training option open. The plan, which requires national service in exchange for tuition costs, would give up to $10,000 per year to those who have graduated from college, twice as much as to those who have only completed high school. This otherwise well-conceived policy reinforces the message of a two-class system. In fact, even though technical training is an explicit part of the program, the media for the most part talk only about college. Once more, technical training, despite the efforts of the secretary of labor and other Cabinet members, remains a residual category.

The three legislators who have long been connected with national service planning—Representative Dave McCurdy and Senators Sam Nunn and Barbara Mikulski—are vociferously against favoring college graduates. Nunn put the argument succinctly: "Skew it to those who've completed college and you've wiped out a major rationale for the plan, which is to get kids of varied backgrounds to work together in a common civic experience. You also want to aid as many members of the college-age population as you can who want decent vocational training. Leave them behind and you pay later in welfare and other costs."[60]

The pitfalls of leaving behind the vocational trainees go well beyond welfare costs. According to the U.S. General Accounting Office, college students are subsidized with federal support amounting to $10,000 per student. Every student who does not matriculate or drops out of school loses the advantage of this $10,000 public investment. This alone is an argument for recalculating the plan to pay equal stipends to college graduates and non-college graduates as the former have already benefited from government support. In addition, the United States has the largest wage gap between the rich and the poor of any industrialized country. If the apprenticeships were successful, many of those young people (and they would not be drawn solely from the poorest socio-economic groups), with their increased knowledge and wage-earning capacities, would help bridge the gap between the poorest groups and the professional elites. They would, in effect, join traditional white-collar workers in creating a revitalized middle class, which is necessary to maintain a functioning democracy.

Politics and Cynicism

Only a committed polity can bring about this fundamental change. Policymakers must be aware of the dangers to democracy if a permanent underclass is created or if the once-solid middle class continues to lose ground. But the public needs the federal government to provide the leadership, funding, and programs to state and local institutions to enable them to carry out the commitment. Public distrust of government has never been greater. Consequently, false promises serve to deepen the cynicism and weaken any national resolve to achieve the admirable goals that have been established.

In 1989, Bill Clinton, then governor of Arkansas, joined President Bush in issuing an education manifesto. Together they assured the country that "by the year 2000, U.S. students will be first in the world in science and mathematics achievement." But there is no way this goal can be achieved by 2000. The data make clear that no one aware of the perilous state of American K-12 education could take this promise seriously. President Clinton has unfortunately reiterated this goal since gaining office, with little to support his continued optimism.

The significant difference between Presidents Bush and Clinton is that the latter has told the country that educational reform will be difficult, expensive, and lengthy, and will require significantly more federal spending, i.e. taxes, which are anathema to the majority of the public. In fact, one of the less heralded impediments to radical restructuring and reinvestment in K-12 education is the parents of school children themselves. In middle and upper-middle class neighborhoods, parents are astoundingly complacent about their children's educations. In careful studies carried out by the University of Michigan five years ago and repeated last year, the authors report that parents continue to rate their local schools highly, believing the problem lies elsewhere, primarily in inner-city schools. In contrast, comparable samples of Taiwanese and Japanese mothers are dissatisfied with their children's educations, no matter how good the schools are. It is very difficult to bring about a revolution in education if a significant part of the populace is complacent even as their children's achievements decline relative to those in other countries.[61]

At the same time, the American public has become increasingly cynical about politics and politicians. Unfulfilled promises deepen cynicism and provoke public unwillingness to make sacrifices for a critical cause. If President Clinton could state forcefully that the only way we will begin to change the direction of the arrow—not to become number one in the foreseeable future, but just to begin to get back on course—is to forget unreal targets and to face the fact that taxes will have to be raised and teachers' training and salaries improved, then he would perform a heroic service to his country. Otherwise, faced with a choice between high skills and low wages, Americans, as presidential advisor Ira Magaziner has warned, are "gradually, silently" choosing low wages.[62]

Innovations

On the positive side of the ledger, many promising new efforts have been launched. In fact, literature and legislation abound with sensible and often inspired plans for bringing American education, training, and the workforce up to the new mark set by a global economy. Individuals and entire organizations within industry and federal, state, and local governments, as well as dedicated teachers and parents, volunteers, professional societies, and private foundations, recognize the depth of the problem and have committed themselves to a radical restructuring of a dire situation.[63]

Without the foundation of a solid K-12 background, all plans for improving the technically-skilled workforce are based on quicksand. Without educating a large part of the public about the demands of education—concentration, deferral of gratification, discipline, and hard work—many good plans will hit an unyielding wall. And without an awareness that innovative plans can bring about unanticipated negative social consequences such as those discussed here—a disparaging distinction between vocational/ technical and college-track programs—some of the most imaginative plans will produce the opposite of their stated altruistic aims.

As a superb communicator, and as someone who overcame family hardship, President Clinton can go beyond the unexciting prose of legislative documents to bring his message to the young people of the country. A recent *Newsweek* poll showed that Clinton's job plan, the hallmark of his economic program, has the support of 57 percent of the public.[64] Clinton's message should convey that he has a plan to make personal futures more secure, and that only by making their futures successful can he bring about the same for the well-being of the country. He should deliver his message relentlessly—"jobs have to be created."

Schools must adapt to the new needs of the individual and society: some college degrees will have to incorporate training for technology, and training will have to continue over a lifetime. And—most important—the invidious distinctions between vocational/technical and college training must be diminished. Only the availability of fulfilling jobs and a narrowing of the gap between the income levels of the college graduates and high school-cum-technical-training graduates can make such a goal anything more than pietistic sermonizing.

The world has changed. What was once the best school system is no longer adequate. Educating just half the population is not enough. Policy makers at national, state, and local levels should initiate public forums on the value of education, work, and the importance of technology in jobs and everyday life. Demonstration programs to publicize examples of successful technology-related college training programs could abet the efforts of cities and towns that are out of the mainstream of major innovations. And many forums should discuss the pros and cons of national service and apprenticeships.

Television, especially MTV, could be the medium of choice for delivering the message. Estimates vary, but the average American child has watched some 5,000 hours of television even before entering school, and by graduation will have watched more than 20,000 hours.[65] The conditions which have led to this mind-numbing experience—the disintegration of the family; the absence not only of a father but increasingly of either parent, as single mothers work longer hours; inadequate daycare; and fewer days spent in school (175–180) than in Western Europe (200 or more) and Japan (220)—are unlikely to be ameliorated in the foreseeable future. Consequently, the business community, already painfully aware of the limitations of the workforce, should engage in a major campaign through television programs and advertisements not only to make learning "cool," but to stimulate an awareness of the central importance of work in a meaningful life.

Unlike boring if well-intentioned public service announcements that go unnoticed by children and young people, announcements could partake of pop culture, cartoons, stroboscopic flashes, and celebrities—the message would be in the medium.

Arsenio Hall led the youth of the country to Bill Clinton, who got their attention by playing jazz on a saxophone. Educators may rail against television, but in front of the TV is where much of the youth of the country, particularly those for whom school is less than a learning experience, can be found.

The reintroduction of the value of work, independent of the years of college education, and the notion that society needs a complex mix of skills and intelligences, will require a bipartisan consensus in Congress, and even more, a public consensus. And it will require the kind of investment that marked the defense buildup during the Cold War.

Twenty years ago, *Work in America* identified the leading causes of discontent in the workforce: "The alienation and disenchantment of blue-collar workers . . . the demands of minorities for equitable participation in 'the system' . . . [and] the search by women for a new identity." Then–Secretary of Health Education and Welfare Eliot Richardson observed in the Introduction that "truly effective responses are far more likely to be made if obscure and complex sources of discontent are sorted out, and the lever of public policy is appropriately placed."[66] Today those "complex sources of discontent" have been sorted out. Now is the time to work the lever of public policy forcefully.

As the landmark report, *A Nation at Risk*, observed in 1983, "If an unfriendly foreign power had attempted to impose on America the mediocre educational performance that exists today, we might well have viewed it as an act of war."[67] Now that the country has redefined national security as economic security, which in turn will be the guarantor of democracy, the battle has indeed been joined with education in the front lines.

Notes

1. President William J. Clinton and Vice President Albert Gore, Jr., *Technology for America's Economic Growth: A New Direction to Build Economic Strength* (Washington, D.C.: Office of Science and Technology Policy, February 22, 1993).

2. I want to thank Norma Paulus, State Superintendent for Public Instruction in Oregon, for providing me with information about the new title and content of innovative high school programs in her state.

3. "Voices from Home And School: Arkansas Parents And Students Talk About Preparing for The World of Work and The Potential for Youth Apprenticeship," A Report on Focus Group Discussions, conducted by Jobs for The Future (48 Grove St., Somerville, Mass. 02144), April 1991. I would like to thank Hilary Pennington, Director of Jobs for the Future, for bringing this and other reports to my attention.

4. Thomas A. Stewart, "Brain Power: How Intellectual Capital is Becoming America's Most Valuable Asset," *Fortune*, June 3, 1991, pp. 44–60.

5. Lynn Williams Interview, "Looking at Workers As An Asset," *Challenges* (Council on Competitiveness) Vol. 6, No. 2 (February 1993), p. 1.

6. Lester Thurow, "Reorganizing Brainpower," *Boston Globe*, October 29, 1991, p. 40.

7. Daniel Yankelovich, editor, *Americans Talk Security*, A Series of Surveys of American Voters, Market Opinion Research, June 1988, p. 34.

8. Ibid., pp. 21 and 37.

9. William J. Crowe, Jr., "Divisive and Peripheral," letters to the editor, *New York Times*, October 13, 1992, p. A22.

10. See Ray Marshall and Marc Tucker, *Thinking for a Living: Education and the Wealth of Nations* (New York: Basic Books, 1992). This work traces the history of the joint effort by Governor Jim Hunt and Dr. David Hamburg, president of the Carnegie Foundation, which led to the formation of the Carnegie Forum and the National Center on Education and the Economy. See *In the National Interest: The Federal Government in the Reform of K-12 Math and Science Education*, a report chaired by Lewis M. Branscomb (New York: Carnegie Commission on Science, Technology, and Government, September 1991). Among many recommendations regarding K-12 education, this report urged the federal government to increase its commitment to aid all students to develop the ability to reason quantitatively.

11. William B. Johnson and A.E. Packer, *Workforce 2000: Work and Workers for the Twenty-first Century* (Indianapolis, Ind.: Hudson Institute, 1987); *The Forgotten Half: Pathways to Success for America's Youth and Young Families* (Washington, D.C.: The William T. Grant Foundation Commission on Work, Family, and Citizenship, 1988); *America's Choice: High Skills or Low Wages!*, Commission on Skills of the American Workforce (Rochester, N.Y.: National Center on Education and the Economy, 1990).

12. William T. Grant Foundation Commission on Youth and America's Future, quoted in Richard Kazis, *Improving the Transition from School to Work in the United States* (Cambridge, Mass.: Jobs for the Future, 1991). I wish to thank Anthony Along, a graduate student at the Kennedy School of Government, who introduced me to the publications of Jobs for the Future.

13. William Celis 3d, "Administration Offers Plan for Better Schools," *New York Times*, April 22, 1993, p. A20.

14. Poll carried out by Public Agenda, reported in their publication, *America's Agenda*, Fall 1992, p. 2.

15. Steven Greenhouse, "Economists Put Forth Plans to Stimulate Growth," *New York Times*, August 31, 1992, p. 12.

16. Peter Costa, "Conversation with Jerome Murphy," *Harvard Gazette*, April 16, 1993, p. 5.

17. See William B. Johnston, "Global Workforce 2000: The New World Labor Market," *Harvard Business Review*, March–April 1991, p. 115.

18. Robert B. Reich, *The Work of Nations: Preparing Ourselves for Twenty-First Century Capitalism* (New York: Alfred A. Knopf, 1991), p. 77.

19. Harvey Brooks, "Educating and Training the U.S. Workforce for the Twenty-First Century" in *The Changing University: How Increased Demand for Scientists and Technology is Transforming Academic Institutions Internationally*, Dorothy S. Zinberg, ed. (Dordrecht, The Netherlands: Kluwer Academic Publishers, 1991).

20. Richard M. Cyert and David C. Mowery, eds., *Technology and Employment: Innovation and Growth in the U.S.Economy* (Washington, D.C.: National Academy Press, 1988), p. 103.

21. Thurow, "Reorganizing Brainpower."

22. For an extended discussion of this topic, see Dorothy S. Zinberg, "Educational Reform and Politics: An Unsatisfying Brew," *Forum*, Winter 1992, pp. 81–82.

23. Study conducted by Comptroller Elizabeth Holtzman, March, 1993.

24. Jonathan Kozol, *Savage Inequalities: Children in America's Schools* (New York: HarperCollins, 1991), p. 69.

25. Census precise figures: 1980, 29.1 percent; 1990, 34.8 percent.

26. "America's Educational System: Still Behind," *Focus* (Ann Arbor, Mich.: National Center for Manufacturing Sciences, August 1992), p. 8.

27. Ibid., p. 7.

28. Jonathan Kozol, *Illiterate America* (New York: Penguin Books, 1986), p. 4.

29. Ibid., p. 9.

30. Harvey Brooks and John Foster, co-chairs, *Mastering a New Role: Shaping Technology Policy for National Economic Performance* (Washington, D.C.: National Academy Press, 1993), pp. 95, 104–105.

31. Survey reported in *The Scientist*, June 8, 1992, quoted in *Manpower Comments*, July–August 1992, p. 6.

32. William R. Grogan, "Engineering's Silent Crisis," *Science*, Vol. 247, No. 4941 (January 26, 1990), p. 381.

33. Barry Cipra, "Math Ph.D.s: Bleak Picture," *Science*, Vol. 252, No. 5005 (April 26, 1991), p. 502.

34. See Dorothy S. Zinberg, "Perspectives on the People's Republic of China: Students and Scholars Abroad Causing Anxiety at Home," *Science*, Vol. 239, No. 4847 (March 25, 1988), p. 1475; and Dorothy S. Zinberg, "Contradictions and Complexity: International Comparisons in the Training of Foreign Scientists and Engineers," pp. 55–88 in Zinberg, *The Changing University*.

35. Johnston and Packer, *Workforce 2000*.

36. Lawrence Mishel and Ray A. Teixeira, *The Myth of The Coming Labor Shortage: Jobs, Skills, and Incomes of America's Workforce 2000* (Washington, D.C.: Economic Policy Institute, 1991).

37. Ibid., p. 3.

38. Ibid., p. 5.

39. David Stern, *Combining School and Work: Options in High Schools and Two-Year Colleges* (Washington, D.C.: Office of Vocational and Adult Education, U.S. Department of Education, March 1991); quoted in Kazis, *Improving the Transition from School to Work in the United States*.

40. Pascal Zachary and Bob Ortega, "Age of Angst: Workplace Revolution Boosts Productivity at Cost of Job Security," *Wall Street Journal*, March 10, 1993, p. 1.

41. Toni Lopeske, "University Systems Suffer under Dwindling Finances," *USA Today*, December 29, 1992, p. 6D.

42. *After the Cold War: Living With Lower Defense Spending*, OTA Report Brief (Washington, D.C.: U.S. Congress, February, 1992).

43. James Medoff, quoted in "Workplace Revolution Boosts Productivity at Cost of Job Security," Zachary and Ortega, "Age of Angst," p. 1.

44. For a prescient discussion of the changes which have become evident in recent years, see Charles A. Zraket, *Understanding and Managing the New Benefits and Problems of the Information Society* (Bedford, Mass.: The Mitre Corporation, July 1982).

45. *The Economist*, August 24, 1991.

46. Pascal Zachary, "White Collar Blues: Like Factory Workers, Professionals Face Loss of Jobs to Foreigners," *Wall Street Journal*, March 17, 1993, p. 1.

47. Ibid., p. A8.

48. John H. Gibbons, "Governing in a Technology-Driven Age: Progress and Problems," *Washington* (Washington University in St. Louis, School of Engineering and Applied Science), October 21, 1991, p. 6.

49. Thurow, "Reorganizing Brainpower."

50. Ibid.

51. I would like to thank Jeff Fine of the Allegheny Trails Council of the Boy Scouts of America, Pittsburgh, who sent me the data from the 1992 Career

Council Report; published in *Focus* (National Center for Manufacturing Sciences), August 1992.

52. Orley Ashenfelter, "How Convincing is the Evidence Linking Education and Income?" delivered as the George Seltzer Distinguished Lecture, Industrial Relations Center, University of Minnesota, 1991. I am grateful to Robert Solow of MIT for providing a copy of this article.

53. Senator William Bradley describing his proposal for "Self Reliance Scholarships" in the *Congressional Record*, July 25, 1991, quoted in Orley Ashenfelter, "How Convincing is the Evidence Linking Education and Income?"

54. Barbara Presley Noble, "Worthy Child-Care Pay Scales," *New York Times*, April 18, 1993, p. F25. The wages earned by the highest-paid childcare center teacher are $15,488 a year, almost $4000 lower than the average salary paid to a woman with only a high school diploma, and $2500 per year lower than the average for a person who takes care of animals in a zoo. As the importance of early exposure to educational experiences has gained recognition (e.g., Head Start programs), the wage scale reveals the society's attitude toward the importance of teachers.

55. Bradley, "Self Reliance Scholarships."

56. Michael Young, *The Rise of The Meritocracy* (London: Thames and Hudson, 1958); p. 32.

57. David T. Kearns and Denis P. Doyle, *Winning the Brain Race: A Bold Plan To Make Our Schools Competitive* (San Francisco: ICS Press, 1988), p. 101.

58. Brooks, "Educating and Training the U.S. Work Force for the Twenty-First Century," in Zinberg, *The Changing University*, pp. 129–140.

59. See Paul Kennedy, *Preparing for The Twenty-first Century* (New York: Random House, 1993); also Kennedy, "Preparing for the Twenty-first Century: Winners and Losers," *The New York Review of Books*, February 11, 1993.

60. Michael Kramer, "Life After High School," *Time Magazine*, March 22, 1993, p. 38.

61. Harold W. Stevenson and James W. Stigler, *The Learning Gap: Why Our Schools Are Failing and What We Can Learn from Japanese and Chinese Education* (New York: Summit Books, 1992).

62. Ira C. Magaziner, "America's Choice: High Skills or Low Wages," in *Software and Hardhats: Technology and Workers in the 21st Century* (Washington, D.C.: Labor Policy Institute, 1992).

63. Among many promising sources see Kazis, *Improving the Transition from School to Work in the United States*; Paul E. Barton, "The School-to-Work Transition," *Issues in Science and Technology* (Vol. 7, No. 3, Spring 1991), pp. 50–54; W.N. Grubb, D. Davis, J. Lum, J. Plihal, and C. Morgaine, *The Cunning Hand, the Cultured Mind: Models for Integrating Vocational and Academic Education* (Berkeley, CA: National Center for Research in Vocational Education, 1990); Jobs for the Future, *Essential Elements of Youth Apprenticeship Programs: A Preliminary Outline* (Cambridge, Mass.: Jobs for the Future, 1991); National Youth Employment

Coalition and The William T. Grant Foundation Commission, *Making Sense of Federal Job Training Policy: 24 Expert Recommendations to Create A Comprehensive and Unified Federal Job Training System* (Washington, D.C.: December, 1992). For an interesting philosophical overview of American education, see "Education: Technical and Moral," in Robert N. Bellah, Richard Madsen, William M. Sullivan, Ann Swidler, and Steven M. Tipton, *The Good Society* (New York: Vintage Books [Random House], 1992).

64. *Newsweek*, April 10, 1993.

65. Paul Kennedy, "The American Prospect," *The New York Review of Books*, March 4, 1993, pp. 42–53.

66. Eliot Richardson, *Work in America*, HEW Commission Report, 1973.

67. The National Commission on Excellence in Education, *A Nation at Risk* (Washington, D.C.: U.S. GPO, 1983), p. 5, quoted in Paul Kennedy, "Preparing for the Twenty-first Century: Winners and Losers," p. 48.

9

Empowering Technology Policy

Lewis M. Branscomb

The U.S. government should put aside its fear of moving beyond the support of science to embrace engineering and technology. We have attempted to provide criteria that can distinguish appropriate modes of support from those that a democratic government in a free-market society should not attempt. Americans cannot afford the luxury of being either so cautious or so overconfident that we fail to earn our place as first among equals in a competitive world.
Alic, et al., *Beyond Spinoff*, p. 410.

President Clinton's assignment of responsibility for scientific and technological affairs to Vice President Gore, their policy statement of February 22, 1993, and the announcement of budgetary plans for technology activities confirm that a high level of attention will be given to technology policy by the new administration. The preceding chapters discuss each of the major components of policy now on the national agenda, identifying both problems and opportunities for addressing them. Through the foresight of Congress in providing a half billion dollars in the FY 1993 defense budget for defense conversion, funds were in hand at the beginning of the new administration to make a significant start on almost all of the initiatives in the new policy. Packaged in the ARPA-led, multiagency Technology Reinvestment Project (see Chapter 3), these funded initiatives have excited intense interest in state capitals and among many others eager to participate.

Daunting obstacles to progress remain, however. A successful national strategy will be both audacious and complex. It must embrace a holistic approach to the generation and use of technical knowledge and the creation of the human resource development and information infrastructure to support these objectives. Success will be highly dependent on government's willingness to place more initiative in the business community and to attract the interest and support of thousands of middle-sized and smaller firms that are the source of much of industrial technology. State governments must be able to concert their actions and provide more continuity in their collaboration with the national government in industrial extension and human resource development.

A number of political difficulties must also be overcome: compensating for distributional effects of success in increasing productivity, avoiding the appearance of conversion of public goods to private gain, resolving the conflicts between the legitimacy and effectiveness of government actions, as well as the conflicts between the call for immediate and visible results and a feasible enabling strategy. Finally, the impatient, risk-averse American political system may be intolerant of the patient, experimental approach, which recognizes the technology-specific character of successful policy, the requirements for institutional innovation, and the need to learn from experience. In the longer term, the missions and even the structure of federal agencies must and will evolve and change.

This chapter describes the requirements for successful implementation of technology policy, then discusses the attendant political difficulties. We then review the suitability of the federal agencies that are the main candidates for implementing the new priorities and policies: the Departments of Defense, Commerce, and Energy, as well as the National Science Foundation. We suggest how federal resources for science and technology might be reallocated if the policies discussed in this book are implemented. Finally, we reiterate our call for an integrated public and private sector approach to creating, adapting, and helping firms use technology to improve the economic performance of the United States.

The Clinton-Gore plan is surprisingly silent on the international dimensions of technology policy. The United States has leveraged its leading position in science to pursue foreign policy goals throughout the Cold War period. Science is still useful for that purpose. Although strengthening the U.S. economy is considered by most to be primarily a domestic issue, technology policy has very important international dimensions.

The highest economic growth rates are not found in the three most highly industrialized regions of the globe—Western Europe, North America, and Japan—but in the developing world and especially in the East Asian "tigers." The potential for further growth is there and also in Eastern Europe and the states of the former Soviet Union. Regional strategies for R&D cooperation must incorporate the business relationships with firms and governments in these new and growing markets. As the United States works at new relationships between government and private industry, it must preserve an open environment for economic and technological alliances across national and regional boundaries. Other economies, especially the newly developing and former socialist countries, must be brought into this vibrant world community of cooperating and competing nations.

The end of bipolarity has also created a new class of countries in severe economic distress, but with high expectations of a better life under capitalism. In common with the underdeveloped nations of the South, the formerly communist nations face a severe lack of capital, of entrepreneurial experience, of infrastructure to sustain a competitive economy, and of the legal, financial, and environmental rules and institutions without which free economies cannot function. But the countries of the former Soviet bloc differ fundamentally from the poor nations of the South. Their people are well educated in science and technology, have a strong national commitment to technology, and have stable population growth. For these reasons, they have the capacity to become "rapidly industrializing," even though their difficulties have not yet peaked. But like the newly industrialized nations of the South, they offer the world the prospect of future growth that the mature industrial democracies cannot expect for themselves.

American economic opportunities in world markets entail expanding foreign investment both by American firms abroad and by foreign firms in the United States. The government's technology policy must balance the desire for international trade with political obligations to national electorates. As new government-industry relationships evolve, how are subsidiaries of foreign firms to be treated? How is international cooperation in science to be managed when the activity is perceived to create commercial advantages for foreign participants? How are international technical assistance programs for helping the former socialist states and the underdeveloped economies of the South to be related to programs aimed at helping U.S. firms compete with highly industrialized nations? Consensus behind public policies to enhance the competitive advantages of American firms and their workers is a prerequisite for resolving these international issues in a politically satisfactory way. Thus a strong domestic policy is required to sustain public confidence in a relatively open world trading system.

Requirements for Successful Policy Implementation

The first-mover advantage that is brought by leadership in introducing new products can be lost to international competitors if the industry is not competitive in making rapid, incremental improvements in product cost and quality. Technological competitiveness requires, therefore, not only the creation of marketable innovations but the ability to commercialize them quickly with superior manufacturing and continue to respond to market changes. Thus the central objective of the policy must be empowering technology for use. This suggests the vital importance of human resources and of institutional settings for realizing applications of both public and private value. This demand-side policy must compensate for market failures, facilitate small-business access to technology, and enable firms to use technology more effectively. A strong demand-side policy helps create the environment for an agile, responsive system for generating new ideas and creating new syntheses of knowledge, the object of supply-side policy. The building of the human resources and information infrastructure will support both. Figure 9-1 suggests how the Clinton-Gore policy initiatives separate into these two categories.

An Integrated Approach to Technology Policy

A technology policy designed to enable private sector capabilities rather than government missions must be organized around private industry rather than government agencies. Such a policy will be decentralized, in the sense that private initiative is encouraged. The goals of public-private partnerships must respond to private opportunities enhanced by public incentives. There must be more

Figure 9-1 Clinton-Gore Technology Initiatives

Supply-side policies: Creating technology	Demand-side and infrastructure: Acquiring and using technology
ARPA dual-use R&D	Industrial extension
Commerce ATP program	Agile manufacturing and enterprise integration
Doubled SBIR grants	
	Information Infrastructure
National Lab research supporting CRADAs with commercial firms	Technology transfer from National Labs to industry
Clean Car project	Workforce training; distant learning
"Green" process technology	
	K-12 education reform
FCCSET Initiatives (e.g., manufacturing engineering)	School-to-Work transition

effective technical linkages between industrial firms and the other sectors of the science and technology enterprise—the universities, national laboratories, training institutions, information service providers.

At the present time most of these linkages are still in their early stages. Since World War II, U.S. technology policy has been integrated around public purposes rather than around private interests. These public purposes were defined around the missions of government agencies. One consequence of this mission-oriented policy is the fragmentation of the science and technology enterprise. Defense industry is isolated from the commercial world. So, too, is the work of the largest national laboratories, whose nuclear weapons work is isolated behind security fences. For most of the postwar period, university research has been only loosely connected to industry research, since the social contract with science was based on an implicit assumption that fundamental research would contribute to the economy without explicit government effort. The nation's *laissez-faire* economic policy discouraged efforts to arrange partnerships between industry and government.

Business firms placed high value on vertical integration, and traditionally kept their suppliers at arms length to gain the advantages of price competition. With strong anti-trust enforcement until the early 1980s, American firms were unaccustomed to collaborating and competing with domestic competitors in order to strengthen their competitiveness in global markets. Industrial firms regarded government with some suspicion, seeing government agencies as a source of increasingly burdensome regulation. All of these characteristics of the U.S. technical environment need to be changed.

In the preceding chapters we have made the case for building the institutional linkages through which new technology diffuses to users, needs for research can be expressed to research institutions, and both people and institutions can be empowered to use technology more effectively. We can identify seven reasons why government's technical activities should be designed to build linkages among firms, universities, national laboratories, and government agencies. Let us consider each in turn.

An Integrated National Industrial Base
The accelerating shift of technology leadership from defense to commercial industry, and diminishing military budgets, compel defense acquisition policy to look to an integrated industrial technology base that will serve both commercial and defense interests. This should take the form of expanded investments in dual-use technology, greater procurement of commercial products adapted to defense use, and helping defense technology suppliers diversify into commercial markets to retain their availability for defense markets.

Linking Universities to Institutions of Innovation
Universities, responding to society's desire for more demonstrable contributions to economic and social well-being, have rapidly expanded their relationships with industry. Questions remain, however, about how the National Science Foundation and the National Institutes of Health should strike the balance between the science culture of academia and the technology culture of industry. In Chapter 7 we suggest how buffer institutions might appropriately bridge the gap in culture, function, and accountability.

New Missions and Delegated Authority for National Laboratories
Similar efforts are being made to reduce the isolation of national laboratories from commercial industry. If the laboratories are given coherent technical missions, and are delegated sufficient authority to manage the delicate balance between those missions and opportunities to collaborate with commercial firms on technical matters of common interest, an important contribution to overall integration of the S&T enterprise is possible. If, however, the laboratory missions fragment and Congress measures their performance by counting CRADAs, the national laboratories will be at serious risk.

Enterprise Integration and Agile Businesses
Commercial firms are rapidly moving away from vertical integration as a means for gaining technological advantage, substituting agile manufacturing and the more collaborative supplier relationships it requires. Corporate alliances allow the acquisition of tacit knowledge much more quickly than earlier attempts to collect all needed skills under a firm's own roof. These trends provide incentive for creating enterprise integration networks, and exploiting the information infrastructure being developed in government policy. The most tightly coupled arrangement of firms is, of course, the consortium that collaborates in technologies of common interest. Government encouragement of industrial cooperation provides for efficient diffusion of technical knowledge, and broadens the distribution of economic benefits beyond a single firm.

Distributed Responsibility for Corporate Technology
Corporate R&D is being decentralized through a great variety of ventures and alliances, both domestic and international. Although there are justifiable concerns about the evidence that the larger American firms are retrenching their levels of R&D investment, this trend may be offset by firms' growing dependence on innovations by their suppliers, which substitutes for in-house capability. Similarly, a growing number of firms see national laboratories and universities as useful sources of commercially relevant knowledge and are learning how to develop the appropriate relationships with them.

Human Resource Development Integrated with Work and Innovation
The public and private sectors must also cooperate in human resource development, with the emphasis on better on-the-job training for workers, on school reform and more effective school-to-work transition, and on new public and private sources of information and education services. As barriers between high schools, vocational education, and college are lowered, respect for skilled work at the pre-professional level can grow. Distance learning and informal education can bridge many of the structural gaps in American education and training.

Industrial Extension Linking All Sectors
Industrial extension services, supplemented by enhanced information infrastructure and human resource development activities, can give innovators better access to existing knowledge and technology, can help firms to find and use technical knowledge and tools, and can motivate the allocation of both public and private resources for new research responsive to identified opportunities for its use. Information networks, both human and electronic, will foster collaboration among all the firms, state institutions, universities, federal agencies, and their national laboratories for all the purposes described above. One often-underestimated benefit that academic scientists and engineers gain from participating in extension services is familiarity with the technical problems of small and large businesses. Their work with small firms enables them to train their students for a broader set of career options, even as they bring their expertise to the service of the firms.

Dependence on Industry Participation

These mixtures of public and private responsibilities call for a higher level of central coordination of the federal agencies, a new relationship between the federal government and the states, and a more effective effort by the federal government to reach out to industry. Soliciting industry initiatives will make technology policy more responsive and less directive. If industrial firms have a strong voice in selection of technology goals and in the construction of the public-private partnerships, they are more likely to respond with the necessary internal technical investments, as well as to provide needed political support. The commitment of industry to the new policies is the critical test for the new programs.

Most of the federal technology programs recognize that industry initiative must be encouraged, but few of these programs are built from this perspective. The Advanced Technology Program (ATP) in the Department of Commerce does rely on unsolicited proposals from industry, and therefore does not limit itself to the government's own technology agenda. But even in that case, a deeper discussion with industry sectors of ATP strategy would undoubtedly give rise to a program with higher leverage and better prospects for positive outcomes.

A Focus for Federal Attention: Small to Mid-sized Firms

The federal government's relationship with the business community varies from tolerant to hostile, except when a particular industry has been singled out by Congress for favorable treatment. The challenge of engaging business support for the administration's technology policy must encounter not only this traditional skepticism of government intentions but also the fact that government must reach out to a segment of the industrial community which has little experience with federal R&D activities: companies that specialize in a specific technology and provide components, materials, or services to larger corporations that assemble parts into finished goods (OEMs). Such firms tend to range in size from the upper half of the small business community (more than 25 employees) up to perhaps 1000 employees. These are the firms identified as candidates for industrial extension services in Chapter 6; they are unlikely to be CRADA partners with national laboratories; they have special needs for human resource development. Typically they do not have research laboratories, but do engage in engineering-based innovations. In the world of agile manufacturing, larger OEM firms depend on them not only for build-to-print parts, but for design engineering and process technology development, thus relieving the OEM client of a part of its R&D responsibility.

These small to mid-sized manufacturers are not well known to government. Unlike the large multi-national OEM firms, they do not have corporate research collaborations with universities, nor do they have political representation in Washington. Unlike the high-tech start-up firms, they are not the main beneficiaries of organized small-business political power. They do not occupy the industrial parks and new-firm incubators that universities are creating in the hopes of replicating Massachusetts' "Route 128," California's "Silicon Valley," or North Carolina's "Research Triangle Park." In Chapter 6, the characteristics of about 50,000 such firms were identified. Reaching them will require the collaboration of state governments and a means for aggregating their interests regionally, or through trade associations.

Success in technology policy is also highly dependent on the right response from state governments. Many new and proposed federal programs require execution by states or at least active cooperation and investment by state governments. The institutional arrangements to make this work well are not in place; and thus federal-state partnerships may prove even more unstable than federal-industry relationships.

Collaboration between the States and the National Government

Most federal agencies do not yet appreciate the extent to which the success of new technology initiatives depends on the active and sustained cooperation of state governments. The states will have to educate them. A recent report of the Carnegie Commission on Science, Technology and Government, prepared by a task force chaired by Richard Celeste, until recently governor of Ohio, identifies the central issue as "how to determine the most effective roles of federal and state government":

Their roles should be developed not on the basis of which level raises (and spends) revenues but according to their relative effectiveness in a given situation, including their effectiveness in catalyzing private-sector action. Determining the appropriate balance in a particular case will require an unprecedented degree of communication and cooperation, with consultation about needs and priorities and timely sharing of information about programs of potential joint interest.[1]

The Carnegie Commission recommended the formation of a new organization to coordinate the states' science and technology activities and to speak for the states in national science and technology councils. While the National Governors' Association's Interstate Working Group on State Initiatives in Applied Research has been active since 1985, and every state participates on its council, the commission's report observes that "the national effectiveness of this organization is limited by its reliance on consensus and its part-time nature. Instead the commission believes that an interstate compact, with a statutory basis in both state and federal law, may "offer the appropriate combination of persistence, independence, and inclusiveness."[2] Such an interstate forum would allow states to negotiate with more consistent voices their collaboration with the emerging federal demand-side technology programs.

Dealing with Political Difficulties

Distributional Effects and Job Creation

Successful technology policy increases productivity, which increases output per unit of labor, but decreases the demand for labor in a static demand situation. This makes selling technology policy as a job-creation strategy a difficult task. The ultimate goal of technology policy is, certainly, increasing the availability and quality of jobs, but the linkage between a successful technology policy and job creation is indirect and delayed. As technology increases productivity, the immediate effect is to reduce the employment needed for a given output. It is only when the wealth created by productivity gains increases demand, through increased purchasing power to workers and lower costs to consumers, that jobs are created to meet this demand. This indirect and delayed relationship between productivity gains and job growth exacerbates the political hazards associated with converting public assets to private earnings.

Technology policy is unavoidably a "trickle down" policy in the sense that firms, rather than individuals, are the first beneficiaries of successful policy. This reality places some fundamental constraints on government action, and strongly suggests that government should restrict its civilian R&D intervention to the classes of market failures that are generally understood and accepted. (See Chapter 3.) Attention to retraining the displaced elements of the workforce will be helpful, especially when technological progress shifts job opportunities from one sector to another. So too is the attempt to provide a more uniform geographic distribution of federal efforts to enhance productivity, so that the successful areas of the country are not perceived as exporting their displaced employment to other less fortunate sections of the country.

Another approach for assuring the broader distribution of benefits is the encouragement of consortia of firms for collaboration with federal civil technology agencies. Such consortia have now become accepted instruments of policy throughout the industrial world. It is doubtful that they have had a decisive effect on economic outcomes in their industries (even in Japan where the MITI-sponsored VLSI project preceded the capture of much of the world's semiconductor memory chip market, and the Fifth Generation Computer Project stimulated the creation of the European ESPRIT program). However, consortia have been a tool for shaping the structure of industry and building new patterns of government-industry relations. MITI has clearly pushed for aggregation of a small number of very strong firms, supported by clusters of specialized suppliers. The European Community has used the Framework Programs and Eureka to change the political and business culture in ways that favor an integrated market for Europe. The United States may be able to use government-promoted R&D cooperation to abate the traditional hostility between government and business and to allay the fears of the public about government taking sides in a competitive market.

Private Gains from Public Funds

The most immediate beneficiary of any strategy to strengthen American economic performance through technology policy is improved firm performance. If this is perceived as the conversion of public assets to private gains, without evident returns to the public good, many people will oppose the policy. As is demonstrated by the political practice of earmarking appropriations for projects in Congressional home districts, government programs that transfer government funds to constituents are more politically rewarding than the more geographically dispersed activities whose outcomes are less tangible. It is hard to compete with targeted programs of high expectations for quick payoff to groups of identifiable voters. But political pressure for more geographically uniform distribution of program benefits will exacerbate the difficulty of a merit-based process of project selection and management.

There are three strategies for containing this political risk. First, government investments in commercially relevant R&D should address demonstrable market failures, as enumerated in Chapter 3. Government must avoid subsidies for commercial product development, and the attendant accusation that the government is simply substituting public funds for private investments that would otherwise have been made. Second, by reaching out to the 50,000 small to mid-sized U.S. manufacturers with industrial extension services, participation in federal technology activities will be more broadly accessible, compensating in some measure for the reduction in defense weapons systems acquisition and manned space projects which helped to stimulate technological modernization in many second-tier manufacturing firms. Third, a partnership with state governments in the administration of demand-side technology activities will help to localize the political context of federal activities. Fourth, federal programs can help firms improve their competitiveness by developing cost-effective technology to meet public goals, such as environmental protection, more equitable health care delivery, and public safety. Such activities have the effect of creating public goods, even as they contribute to the economy.

Conflict between Legitimacy and Effectiveness

This strategy, however, has its own pitfalls. When government sets out to enhance the competitiveness of commercial firms, it is seeking to accelerate progress toward goals that free citizens pursue in a capitalist economy. Assuring a healthy economy is certainly an appropriate concern of government. But both social activists and libertarians might agree that in the economic sphere a primary responsibility of government is to compensate for the undesirable externalities of the market. Policy tools for these ends tend to be regulatory or redistributional and may be antithetical to efforts to accelerate economic growth. Technology policies are often expected to bridge this gap. Consider two examples: first, in the Clean Car project discussed in Chapter 3, the objective is enhanced economic performance of the automobile industry, but the chosen technology directions contribute to pollution abatement, increased safety, and energy import avoidance. The program gains legitimacy from these public interest goals, but they may not be effective in producing lower cost, better functioning automobiles. Second, in the case of the NIH Cooperative Research and Development Agreements discussed in Chapter 4, the government set in place strong incentives for the transfer of biomedical inventions from NIH laboratories to industrial firms, and when these CRADAs begin to look successful, members of Congress begin to ask NIH to require price constraints by participating firms as a condition of their access to government scientific results.[3] These pressures legitimate the use of public funds in the CRADA work, but may be obstacles to the cooperation for which the CRADAs were established.

There is no ideal solution to this conflict between legitimacy and effectiveness, but political risks might be mitigated by restraining government investments to R&D in cases of demonstrable market failures or other well-understood reasons for private sector underinvestment in relation to returns to the nation as a whole. (See Chapter 3.)

Conflict between Expectations for Results and an Enabling, Demand-side Strategy

How can the administration build a constituency for public investments in civilian industrial technology? After forty years of federal R&D activities aimed at producing new weapons, conducting space missions, or discovering new disease therapies, the public expects government R&D programs to produce visible outcomes. The most conspicuous evidence for the political compulsion for demonstrable consequences from public technology expenditures is the "megaproject" to demonstrate the commercial potential of a new technology. As Cohen and Noll observed, the political appeal of these demonstration projects is so great that they are in danger of becoming captured by their constituencies, failing to adapt to changing technology or market conditions, and then failing to meet their initial goals.[4] We are recommending a very different kind of national technology strategy, one that does not preempt private industry's priorities, but instead is intended to enable private capabilities to pursue goals of their own choosing.

An Experimental, Flexible Approach

The administration of these new policies will not be easy. Most of the civilian agencies lack experience with investing in the industrial technology base and in industrial extension and information infrastructure; these programs can only grow with experience. With the locus of action shifting to the private sector, the technical agenda grows correspondingly broader than the government's historic technical missions. General policies that have been effective in one industry may be quite inappropriate for another. Government officials must be industrially experienced and must recognize the idiosyncratic nature of specific technologies and of each particular high-tech industry. Government officials will have to become much more sophisticated in technology, economics, and politics than was necessary to administer basic research support to universities and manage the government's own technological responsibilities.

Officials newly in office feel urgent pressures to make very rapid progress toward promised goals. They may find it hard to accept the necessity of incremental institutional and program learning. But virtually every one of the technology policy initiatives needs to be handled that way. There are many reasons for this advice: first, as noted above, success requires that participating firms gain confidence in the government as a partner and become willing to support the joint technology efforts politically and with their own internal investments. This confidence must be earned through successful experiences. Second, the specific criteria for managing these technical activities are technology-specific; government agencies must acquire the skills and experience appropriate to each industry they work with, and must recognize their many differences—in market structures, the importance of intellectual property protection, the sources of technical support, sensitivity to foreign trade, etc. Third, each of these programs is dependent on the others, as discussed above. In addition, most of them require development in cooperation with fifty state governments, all of which requires time to develop. Finally, Congress is understandably reluctant to increase appropriations for any agency faster than 20 to 30 percent per year, regardless of its size. Most of the technology policy innovations aimed at industrial competitiveness are being constructed on very small base levels. NIST's Advanced Technology Program, for example, began in 1989 with a $9 million appropriation, and in spite of being doubled almost every year since, is in FY 93 still below $70 million.

The need for an experimental approach will test the public's patience and the government's perseverance. It will also test the willingness of Congress to support the measurements and analysis required to evaluate the effect of the agencies' efforts.

Patience and Perseverance in Public-Private Partnerships

Government must find a way to commit resources to the patient development of collaborations with private institutions and state governments. Provision for multi-year appropriations at more predictable levels is needed to overcome the difficulties of such collaborations. In almost all of the activities of the new technology policy agenda there are complex mixtures of public and private responsibilities. Many of these involve the creation of new forms of inter-institutional relationships, such as the National Research and Education Network, University-Industry Research Centers, and industrial extension services that combine both public and private resources. Others call for new relationships, such as national laboratory partnerships with industry and the acquisition of commercially developed technology for military use. Each of these examples involves collaboration among institutions whose cultures are to some degree incompatible.

Even within a single firm, each of the different levels of technical activity (fundamental research, product development, or production engineering, for example) performs best in the environment best suited to its needs. The push for enhanced competitiveness requires that each level work more closely and more concurrently with all the others. If the coupling is too strong, the special environment needed for basic research or for market-responsive innovation may be eroded; if it is too weak, technical capabilities will be underutilized. This balance is hard to strike under the best of circumstances. It is particularly difficult across a government-business culture gap.

Political support for demand-side policy must be institutionalized in much the same way that cooperative extension has been in agriculture, rather than depending on the narrow and deep constituencies we associate with the megaprojects of mission-driven policy.

Government Organization and Priorities

If all of the political and policy obstacles outlined above are overcome, how suitable are the structure, missions and experience of the federal agencies that are being called on to implement a new technology policy and rapidly changing national R&D priorities? As noted in Chapter 3, Americans are quite pragmatic about institutional innovation and tend to prefer adapting existing institutions to new policies rather than trying to restructure the government. Crisis certainly increases the institutional mutation rate, but evolution is more likely than top-to-bottom reorganization. Wise government policy makers will attempt to assign new responsibilities to those departments and agencies that are in the most favored position to get Congressional support to develop the capability to perform the new tasks. Competing political and economic interests will, of course, favor the agencies on which each has the best hold. What agencies are most likely to emerge from this competition?

There are three primary candidates in the federal government for the major role in creating and enabling civilian technologies of economic importance: the Department of Defense, and especially its dual-use technology agency, ARPA; the Department of Commerce, its Technology Administration and its flagship agency NIST; and the Department of Energy, with its massive national laboratories. Another important player will be the National Science Foundation; although dedicated primarily to support of university research, NSF has important roles in all fields of science and engineering research, in technology diffusion, and in education and training, and has played a lead role in developing the Internet on which the information infrastructure strategy is built. The NIH, with a substantially larger budget than NSF, will continue to dominate the health sciences and their support for the emerging biotechnology industry. Except for the leadership shown by the National Library of Medicine in the information infrastructure, however, NIH has not attempted to exert significant influence outside biology and the health sciences and their associated industries. The National Aeronautics and Space Administration (NASA) is too narrowly focused and too constrained in mission and budget to be a major player, except in the field of aeronautics.

The most likely outcome is that all of these agencies will be promoted by proponents in Congress, and the structure of government will look in ten years much as it does today. To remain a major factor in the execution of technology policy, however, each of these departments—Defense, Commerce, and Energy—will have to change its culture and its policies substantially. Defense will have to give up its unilateral determination of technical requirements and R&D strategies if it is to gain better access to commercial technology. Commerce will have to learn to be more agile and innovative and transcend its intramural culture as it builds up the size of its support of industrial R&D. Energy will have to find new and compelling missions that are a good fit to the structure and talents of the laboratories, and delegate more decision authority to those laboratories cooperating with commercial industry. Let us consider each of these agencies in a little more detail.

The Department of Defense

Defense conversion is the policy basis for a major role by ARPA in the Department of Defense. Assisting military contractors to convert their activities from declining defense work to expanded activity in commercial markets might seem a temporary, transitional need. But the expected shift of federal resources from defense to civil activities is planned to take place over a number of years. Perhaps more important, the objective of conversion is the creation of a unified industrial technology base, supporting both commercial markets and defense acquisition.[5] The Technology Reinvestment Project is much more sophisticated than the efforts after the Vietnam War, when military aircraft producers turned to aluminum canoes and light-rail cars. As more "spin-on" technology from commercial firms is applied to military applications, the one-directional nature of technology flows we associate with defense conversion today will become bi-directional. In that form it will be a permanent state of affairs. ARPA, and perhaps other defense agencies, will have sufficient R&D resources—even in a substantially attenuated Department of Defense—to be a major player in the co-development of technologies for both military and commercial use. DARPA was renamed ARPA in March 1993 specifically in anticipation of this role.

Is the new role for ARPA, symbolized by the Technology Reinvestment Project, a transitional affair, or will ARPA continue to play a leading role in civilian technology policy under the cover of defense conversion? The "defense conversion" rubric is likely to evolve, as a reflection of American pragmatism in institutional innovation, into a continuing responsibility for significant investments in dual-use technology. The reform of the military acquisition system is likely to be very slow and difficult. The notion of defense conversion is likely to change from the image of firms converting swords to plowshares, to the image of the Department of Defense converting from a command economy dealing with a captive defense industry, to an agency negotiating technology sharing between commercial and military products through a broad variety of alliances and joint ventures. There may never come a time when the process of adjustment is "over."

The Department of Commerce

By name, mission, and constituency, the Commerce Department is a logical candidate for the competitiveness enhancement mantle. As discussed in Chapter 3, NIST is the most experienced agency in collaborating with the broadest spectrum of commercial industrial firms. Its new Technology Administration has the Congressional charter to win for itself a primary role in trade-related issues, in industrial extension services, and in co-funding with private firms pre-competitive research and development through the Advanced Technology Program. This department's primary disadvantage in competing with Defense and Energy is its relatively small budgetary base in research and development, and doubts about its ability to build a strong political constituency. However, this second handicap might, in the long term, prove a major advantage.

There is little basis for Commerce receiving large appropriations for R&D in support of the commercial industrial base unless its constituency—the business community—wants it and works for it. In the past the business community has been highly skeptical of government efforts to help it compete, although there are signs that this is changing, as noted in Chapter 6. If the business community gets behind the government program, helps to make it effective, and supports it in Congress, the ability of the Commerce Department to sustain a significant role in industrial competitiveness would be greatly strengthened. In any case, just as the Department of Defense may evolve toward the role of partner and customer of an integrated commercial-military industry, the Department of Commerce will struggle to evolve, no doubt slowly, toward the *de facto* role of a "Department of Industry and Technology."

Organizationally, the department needs to build up its capabilities using NIST as the primary source of competence in industrial technology. NIST's predecessor agency, the National Bureau of Standards, often demonstrated its ability to nurture and then spin off agencies to meet new challenges, particularly during World Wars I and II and during the Korean War.[6] NIST should again be split into two agencies, one continuing the excellent work of the NIST research laboratories and the other structured to manage the extramural programs: the Advanced Technology Program, industrial extension, and the manufacturing technology centers. The first of these might be called the National Laboratory for Standards and Technology, and the second called the Industrial Technology Agency. Both would report to the Undersecretary for Technology, along with the National Technical Information Service, the National Telecommunications and Information Agency, and the

Patent and Trademark Office, all of which should be brought into the Technology Administration.

The Department of Energy

The Department of Energy is in many ways the most exposed to change of the three departments, for the driving force for this extraordinarily R&D-intensive department has been its National Defense Program, building nuclear weapons, and the Naval Reactors Program, building nuclear powered submarines. If the nuclear arms control treaties with the states of the former Soviet Union are signed and fully implemented, the core mission of the department will be deeply affected. As discussed in Chapter 4, a substantial part of the annual expenditures of $6.6 billion in the DOE laboratories will be subject to cuts.[7] Those who want to keep this massive and quite broadly-based technical capability at full strength are looking for a new mission that can command a high priority in the federal budget. Senator Bennett Johnston and Senator Pete Domenici, the authors of proposed legislation (S. 4), see DOE's mission as technology partnerships with industrial firms, applying the technical skills and embedded knowledge of the DOE laboratories to a broad range of commercial activities.

If this effort is successful, the Department of Energy will have taken a small but significant step toward the functional role of a "Department of Science and Technology" recommended in the report of President Reagan's Commission on Competitiveness.[8] There is, however, widespread opposition to the idea of aggregating much of the federal R&D effort into one large, broadly missioned department. Some fear it makes R&D a tempting target for budget cutting or political earmarking. Others make the argument that the United States has benefitted greatly from its decentralized responsibility for R&D. Each agency justifies its R&D expenditures in terms of its mission, and trades off those expenditures against alternative means for doing the job. Thus there are many justifications for R&D and there is no cap on the R&D total in the federal budget. R&D has to pay its way by demonstrating its value in the achievement of operational goals and capabilities.

There is an alternative direction in which the Department of Energy might evolve that offers more promise. The Department of

Energy might build on its existing activities, especially the more technological aspects, in environmental remediation and in new energy sources and enhanced efficiency. At some future date, the Department of Energy might become the Department of Energy and Environment through a merger with the Environmental Protection Agency (EPA) and acquisition of the National Oceanographic and Atmospheric Administration from the Department of Commerce and of the Geological Survey from the Department of Interior.[9] This would facilitate the redeployment of the Energy Department's laboratory capability to the search for environmentally superior as well as more energy-efficient technologies. The Clinton administration has embraced the idea of government investments, cost-shared with industry, in environmentally superior process technology, but EPA has little capability to manage this kind of technology development activity. The merger of Energy and EPA would further facilitate tradeoffs between energy policy and environmental goals; while DOE and EPA are in separate institutions and look to separate constituencies, these tradeoffs are hard to make. Finally, the merger would permit a net reduction in administrative costs by reducing the number of independent units reporting to the president.

The National Science Board and National Science Foundation

The National Science Board (NSB), which sets policy for the National Science Foundation (NSF) and authorizes its grants, is trying to find the right balance between its traditional focus on support for fundamental academic research and the growing expectation that it will embrace increased responsibility for the commercialization of that research. This choice was, in effect, the issue facing the National Science Board Commission on the Future of NSF, described in Chapter 7. Many of the mechanisms through which science creates public benefits are beyond NSF's control; they are managed by federal agencies other than NSF or are conducted by firms, professional societies, and other non-governmental organizations. If the NSB chooses to take a passive stance on the commercialization of science, it runs a serious risk that political frustration with the processes linking science to public benefit will erode support for basic science and diminish the role of NSF. This leaves NSF with two choices: either engage its programs in the entire technological "food chain" from research to market, or take an active role in helping the president and his science advisor formulate a technology policy within which NSF's proper role is defined and responsibility for the linkages between science and public benefits is assigned in a way satisfactory to NSF. The first course would be strongly opposed by the universities; the second is NSB's responsibility and is consistent with its legislative mandate, which has hitherto been little exercised.

The NSF should continue to support research in the sciences and in engineering, and support the facilities and infrastructure required for a strong national research capability. NSF should not tell university scientists what to do; it need only put properly allocated resources in the paths of bright people. The present portfolio of NSF activities is broad enough, but NSF should follow the advice of its Commission on the Future of NSF and consider how it might improve the process of resource allocation to the many areas of science and technology in which it invests. In this process advice from the most technically qualified people with industry experience should be sought. NSF should give increased attention to fields of science and engineering research that could make a very big contribution to the nation's technological capability in the long term. As discussed in Chapter 7, technological problems are often a source of new and challenging problems in basic science. NSF should pay more attention to disciplines and interdisciplinary research known to be underrepresented in American universities. Rather than engaging in opportunistic problem-solving (applied) research, however, NSF should carry technology-driven research deeper and with greater generality than industrial laboratories are likely to do. It should search for deeper understanding of principles underlying interesting technologies than is required to solve the immediate problems the technology addresses.

Resource Allocation

The debate about government investment in technical activity ultimately must answer the question: how much is enough? As

Defense Department expenditures on R&D and on procurement (which drives R&D in supplier firms) decline, how will the resources thus freed up be reallocated? President Clinton wants to keep federal R&D expenditures at current or higher levels, but he has declared his intention to shift the balance of military to civilian R&D from equality to forty percent military and sixty percent civilian. This entails the redeployment of nearly $8 billion of federal funds to civilian R&D. This massive reprogramming of government R&D will be difficult to accomplish for two reasons. It is always politically difficult to shift appropriations from one Congressional subcommittee's jurisdiction to another's. Second, the civilian technology programs to which the funds might be reprogrammed are starting from a much lower base than the much larger military programs.

However, even if this shift in the federal R&D budget is attained, there remains the need to increase the ratio of private sector R&D to that of the public sector. In other words, some of the downsizing in defense should probably reappear as private investment, especially if, for the reasons given, it is hard to make the budget shift the administration intends. But the desired growth of private sector R&D may also be difficult to achieve. Commercial innovation is substantially less R&D intensive than military innovation, by a factor of four.[10] If the administration were to reduce the acquisition of military products by $8 billion and use the funds to stimulate commercial production, at traditional levels of commercial R&D intensity, the same level of commercial production would demand only one-fourth as much R&D to support it. If, on the other hand, money for military R&D is shifted to civilian R&D, and this R&D supports commercial production in the normal ratio of R&D to sales, a four-fold increase in economic activity would be required to utilize the newly funded civilian R&D. This line of argument, combined with the expected growth of dual-use technology support by defense agencies (see the next section), suggests that the shift from defense to government civil programs and to private R&D will not be as quick as President Clinton promised during the 1992 campaign and in the early months of his administration.[11] There are, of course, many civil activities, both public and private, that would benefit from becoming more R&D intensive.[12]

Other parts of the government's R&D budget seem likely to shrink as well. The Department of Energy laboratories seem certain to shrink with the reduction in nuclear weapons activities, even as they expand their collaborations with industry and enter new areas such as environmental and transportation research. The last time there was a broad reduction in military activity and a shift to a civilian agenda was in the early 1970s, as the Vietnam War was winding down. At that time the rapid growth of very expensive commercialization demonstrations, mostly for new energy sources, helped take up the slack from declining defense budgets. But the effectiveness of such demonstrations has been generally discredited.[13] Most of them had been terminated by the first year of the Reagan administration, and they are unlikely to be resuscitated. The new administration is also showing restraint in funding the space station and the Superconducting Supercollider, and Congress, feeling the pressure to reduce the deficit, is increasingly skeptical of these technical megaprojects. Programs implementing demand-side policies generally call for less federal R&D investment than does supply-side technology creation. For all these reasons, a reduction in total federal R&D expenditures in the next three or four years is very likely. The open question is whether the restructuring of American industry and a resurgent economy can provide enough incentive for expansion of private R&D investment to balance the loss of government effort and to give the United States a public-private investment balance a little closer to that of Japan (21.5 percent public and 78.5 percent private in 1988).[14]

There is one more unanswered question about the right balance between publicly funded civil R&D and private investment. As noted in Chapter 7, technological innovation has become more "scientific" with the result that the upstream end of the innovation process has tended to become more of a public good. Scientific knowledge is beginning to form a common base of knowledge from which everyone competes, rather than the primary source of competitive advantage. Competitive advantage arises more from elements further along the technological "food chain": manufacturing processes, product and service quality, complementary assets, and quickness of response to market opportunity. This does not imply a reduction in that amount of scientific research supporting industrial innovation, only that the upstream investment would receive increasing public support and more private support through cost sharing by firms in consortia or alliances. Just these trends are reflected in the technology policy.[15]

Of course, there is no magic number for the "right" national R&D investment. The shortcomings of comparing the R&D intensity of one economy with another have been discussed above. It is our conviction that the recommendations in this book, if successfully implemented, have the potential to increase the effectiveness of both public and private R&D so that even a reduced level of total effort could make a substantially larger contribution to the economic well-being of society. But the supply of technically skilled human resources is not perfectly elastic, nor is the labor market for scientists and engineers perfectly efficient in allocating the best skills to the most urgent and demanding tasks. When industrial demand for scientists and engineers falls and large numbers of technical people in national and government laboratories are looking for new jobs, universities find it increasingly difficult to attract students into science and to place those who are finishing their studies. Thus it is doubly important that the government be successful in its efforts to build the public-private partnerships envisioned here, and thus to ameliorate the severe dislocations in the science and technology enterprise that are now taking place. But the nation should also have in mind a long-term goal for a more appropriate balance of resources among the major institutional players in science and technology.

An example of what such a new balance might look like is shown in Figure 9-2. No analytical methodology can yet substantiate that the allocations illustrated in this table are economically optimal or are politically achievable. They are only intended to express the directional consequences of the policy recommendations in this book.

Americans now appreciate that the government's traditional agenda is a poor match to the technical challenges facing American manufacturers. Even when government projects do create technology of potential value to commercial firms, we have learned that spinoff from government R&D is neither automatic nor free. On the other hand private industry, left to its own devices in a *laissez-*

Figure 9-2 Resource Allocation Implications of Policy Change

Source of funds	1992 distribution of nat'l R&D expense	Possible distribution in year 2000
Private industry	51.5 percent	60 percent
Federal resources	43.5 percent	35 percent
States, universities, nonprofits	5.0 percent	5 percent

Performer shares of federal funds	1992 expense (percent)	Possible distribution in year 2000
National and government labs	13.5 percent	8 percent
Federal support to universities	6.4 percent	8 percent
Federal support to buffer institutions at universities	0.5 percent (est.)	3 percent
Federal R&D support to civilian firms & consortia	0.8 percent (est.)	4 percent
Federal support for defense industry and dual-use R&D	20.5 percent	8 percent
Federal support to not-for-profit institution	1.8 percent	2 percent
Federal information infra-structure and extension services	< 1 percent	2 percent
Federal education R&D	< 1 percent	1 percent

Data from National Science Board, *Science & Engineering Indicators—1991*, NSB 91-1 (Washington, D.C.: U.S. Government Printing Office, 1991), Table 4-2, p. 307. Percentage for buffer institutions is an estimate, subtracted from funds to universities. Percentage for consortia is an estimate, subtracted from funds to industry.

faire economy, will seriously underinvest in many areas of research that would pay large returns to society as a whole. Nor will markets act to create the supportive infrastructure needed to support a high value-added economy and a well-trained workforce. The correct role for government is to pursue a demand-side policy that helps all technical institutions understand what their technical opportunities are, where to find and acquire the technology they need, and how to get it created when necessary. Through such a strategy, government can empower technology without controlling it. The confidence that firms and governmental institutions would gain in a shared vision of a knowledge-intensive society could create the demand for new knowledge and higher skills that can make the vision a reality.

American scientists and engineers have set new standards for scientific excellence in many fields. Many scientists are worried that their fellow citizens are losing confidence in the power of science to better their lot. They worry that if other nations are seen as exploiting American science more effectively than our own industry, the government will fail to invest in sustaining the leading position American science still enjoys. The capability-enhancing policies we recommend are designed to ensure that this does not happen. Placing more emphasis on the linkages, services, and human resources that better exploit that scientific leadership may be the best strategy for rekindling the enthusiasm for long-term investments in fundamental research and creative invention in America.

Notes

1. Carnegie Commission on Science, Technology and Government, *Science, Technology, and the States in America's Third Century* (New York: Carnegie Commission, September 30, 1992), p. 10.

2. Ibid., p. 28.

3. Congress, Office of Technology Assessment, *Pharmaceutical R&D: Costs, Risks, and Rewards*, OTA-H-522 (Washington D.C.: U.S. Government Printing Office, February 1993), pp. 35–37 in summary. See also discussion of this issue in Chapter 4 and Chapter 7, and references there.

4. Linda Cohen and Roger Noll, *Technology Pork Barrel* (Washington, D.C.: The Brookings Institution, 1991).

5. John A. Alic, Lewis M. Branscomb, Harvey Brooks, Ashton B. Carter, and Gerald L. Epstein, *Beyond Spinoff: Military and Commercial Technologies in a Changing World* (Boston: Harvard Business School Press, April 1992).

6. The National Bureau of Standards (NBS) helped create an optical glass industry in World War I, and a synthetic rubber industry during World War II, when it also spun off its proximity fuse work to form the Diamond Ordnance Fuse Laboratory. Later it created and spun off the China Lake Laboratory for the Navy. Its ionosopheric propagation laboratories became the nucleus for the agency that evolved into the National Oceanographic and Atmospheric Administration.

7. The Clinton administration budget for FY94 includes substantial increases for ARPA and NIST and a decrease of a few percent in Department of Energy R&D.

8. President's Commission on Industrial Competitiveness, John A. Young, chairman, *Global Competition: the New Reality*, Vols. 1 and 2 (Washington, D.C.: U.S. Government Printing Office, 1985).

9. The decision of the administration to ask Congress to elevate EPA to the status of Department of the Environment makes such a merger quite unlikely for some time.

10. Alic, et al., *Beyond Spinoff*, p. 167.

11. I am indebted to Harvey Brooks for pointing out this dilemma arising from the great differences in R&D intensity between civilian and military innovation.

12. Harvey Brooks, "The Future: Steady State or New Challenges?" in Susan E. Cozzens, Peter Healey, Arie Rip, and John Ziman, eds., *The Research System in Transition*, Nato ASI Series D, Behavioral and Social Sciences, Vol. 57 (Dordrecht: Kluwer Academic Publishers, 1990), pp. 163–172.

13. Cohen and Noll, *Technology Pork Barrel.*

14. Alic, et al., *Beyond Spinoff*, Table 7-1, p. 211.

15. This observation is from Harvey Brooks.

Acronyms

AAAS	American Association for the Advancement of Science
ACDA	Arms Control and Disarmament Agency
ACTA	Advanced Civilian Technology Agency
AID	Agency for International Development
ANS	Advanced Network and Services, Inc. (provides NSFNET backbone services)
APS	American Physical Society
ARPA	Advanced Research Projects Agency
ARPANET	Advanced Research Projects Agency Network
ATP	Advanced Technology Program (at NIST)
AT&T	American Telephone and Telegraph Co.
AUP	Acceptable Use Policy (NSF guidelines for traffic on the NSFNET backbone network)
BEST	Board on Environmental Sciences and Toxicology (NAS)
BLS	Bureau of Labor Statistics
CAD	Computer-Aided Design
CAE	Computer-Aided Engineering
CALS	Computer-aided Acquisition and Logistics Support (DOD)
CAM	Computer Aided Manufacturing
CARP	Civil Automotive Research Program (1979)
CARPA	Civilian Advanced Research Projects Agency
CAT	Computer Aided Tomography (CAT scan medical device)
CCL	Controlled Commodities List (Dept. of Commerce export control list)

CEA	Council of Economic Advisers (in Exec. Office of the President)
CEO	Chief Executive Officer (of a firm)
CIM	Computer Integrated Manufacturing
CIM	Corporate Information Management (DoD)
CIX	Commercial Internet eXchange
CNC	Computer Numerical Control
CNRS	Centre Nationale de Research Scientifique
COCOM	Coordinating Committee (for export control)
COSEPUP	Committee on Science, Engineering, and Public Policy (NAS/NAE/IOM)
CRADA	Cooperative Research and Development Agreement
CSIRO	Commonwealth Scientific and Industrial Research Organization
CSPP	Computer Systems Policy Project
CTI	Critical Technologies Institute
CTC	Civilian Technology Corporation
CTO	Chief Technical Officer (of a firm)
DARPA	Defense Advanced Research Projects Agency
DDR&E	Director of Defense Research and Engineering
DELTA	Department of Energy Laboratory Technology Act
DNA	Deoxyribonucleic Acid
DOC	Department of Commerce
DoD	Department of Defense
DOE	Department of Energy
DOT	Department of Transportation
DTCC	Defense Technology Conversion Council
EC	European Community
EINet	Enterprise Integration Network (MCC)
EOP	Executive Office of the President
EPA	Environmental Protection Agency
ERC	Engineering Research Center (NSF)
ESPRIT	European Strategic Program for Research and Development in Information Technology
Eureka	Not an acronym; refers to an EC program to encourage transnational commercial collaboration
FAA	Federal Aviation Agency
FACA	Federal Advisory Committee Act

FCCSET	Federal Coordinating Committee for Science, Engineering and Technology
FFRDC	Federally Funded Research and Development Center
FNC	Federal Networking Council
FOIA	Freedom of Information Act
FY	Fiscal Year
G&A	General and Administrative (expenses)
GAO	General Accounting Office
GDP	Gross Domestic Product
GE	General Electric Corp.
GOCO	Government-owned Contractor-operated (laboratory)
GOGO	Government-owned Government-operated (laboratory)
GPO	Government Printing Office
GUIR	Government-University-Industry Research Roundtable
HASC	House Armed Services Committee
HHS	Health and Human Services Department
HPCC	High Performance Computing and Communications
HPCCIT	High Performance Computing and Communications and Information Technology subcommittee (FCCSET/PMES)
HUD	Housing and Urban Development Department
IAB	Internet Architecture Board (within the Internet Society)
IBM	International Business Machines Co.
IETF	Internet Engineering Task Force (IAB)
IINREN	Interagency Interim NREN
IOM	Institute of Medicine
IP	Internet Protocol
IR&D	Independent Research and Development
ISDN	Integrated Services Digital Network
ITES	Industrial Technology Extension Service (New York State)
IVHS	Intelligent Vehicle, Highway System
MCC	Microelectronics and Computer Technology Corporation
MCTL	Military Critical Technologies List
MITI	Ministry of International Trade and Industry (of Japan)
MOC	Manufacturing Outreach Center
MTC	Manufacturing Technology Center
NAE	National Academy of Engineering

NAP	Network Access Point
NAS	National Academy of Sciences
NASP	National Aerospace Plane
NEC	National Economic Council
NEMTC	Northeast Manufacturing Technology Center
NII	National Information Infrastructure
NIST	National Institute for Standards and Technology (Dept. of Commerce)
NOAA	National Oceanographic and Atmospheric Administration (Dept. of Commerce)
NREN	National Research and Education Network
NRC	National Research Council (of NAS, NAE and the IOM)
NRL	Naval Research Laboratory
NSB	National Science Board
NSC	National Security Council
NSF	National Science Foundation
NSFNET	National Science Foundation Network (backbone for research and education; part of the Internet)
NTIA	National Telecommunications and Information Administration (Commerce Department)
NTIS	National Technical Information Service (Commerce Department)
OECD	Organization for Economic Cooperation and Development
OEM	Original Equipment Manufacturer
OMB	Office of Management and Budget
OST	Office of Science and Technology (predecessor to OSTP)
OSTP	Office of Science and Technology Policy
OTA	Office of Technology Assessment (U.S. Congress)
PCAST	President's Council of Advisors on Science and Technology (to President Bush)
PCS	Personal Communications Services (digital, wireless service for voice and data)
PDES	Product Data Exchange using STEP
PENNTAP	Pennsylvania State University's Technology Assistance Program
PL	Public Law
PMES	Committee on Physical, Mathematical and Engineering Sciences (FCCSET)
PSAC	President's Science Advisory Committee (from Eisenhower to Nixon)

PTFP	Public Telecommunications Facilities Program (NTIA)
R&D	Research and Development
RBOC	Regional Bell operating company
RCA	Radio Corporation of America
RDT&E	Research, Development, Test and Engineering (Defense Dept. budget category)
S&T	Science and Technology
SBIR	Small Business Innovation Research
SDI	Strategic Defense Inititiative
SEAB	Secretary of Energy Advisory Board (re: national labs)
SemaTech	Semiconductor Manufacturing Technology
SIC	Standard Industrial Classification
SLAC	Stanford Linear Accelerator
SRI	Stanford Research International
SSC	Superconducting Supercollider
S&T	Science and technology
START	Strategic Arms Reduction Treaty
STEP	Standard for Exchange of Product Model Data
STEP	State Technology Extension Program (NIST)
STS	Science Technology and Society
T&E	Test and Engineering
TAAC	Technology Assessment Advisory Council
TCP/IP	Transmission Control Protocol/Internet Protocol (communications protocols used on the Internet)
TECNET	Technology for Effective Cooperation Network
TQM	Total Quality Management
TRP	Technology Reinvestment Program
UIRC	University-Industry Research Center
USPTO	United States Patent and Trademark Office (DOC)
VLSI	Very Large Scale Integration (of microelectronics)

Bibliography

Science Policy and Its Origins

Vannevar Bush, *Science—The Endless Frontier,* A Report to the President on a Program for Postwar Scientific Research (Washington, D.C.: Office of Scientific Research and Development, July 1945; repr. National Science Foundation, 1960, 1980, 1990), pp. 5–40

Yaron Ezrahi, *The Descent of Icarus: Science and the Transformation of Contemporary Democracy* (Cambridge: Harvard University Press, 1990)

Bruce L.R. Smith, *American Science Policy Since World War II* (Washington, D.C.: Brookings Institution, 1990)

John M. Ziman, *Public Knowledge: An Essay Concerning the Social Dimensions of Science* (Cambridge, England: Cambridge University Press, 1968)

Contemporary Discussions of Technology Policy

John A. Alic, Lewis M. Branscomb, Harvey Brooks, Ashton B. Carter, and Gerald L. Epstein, *Beyond Spinoff: Military and Commercial Technologies in a Changing World* (Boston: Harvard Business School Press, April 1992)

Lewis M. Branscomb, "Does America Need a Technology Policy?" *Harvard Business Review,* March–April 1992, pp. 24–31

Lewis M. Branscomb, "America's Emerging Technology Policy," *Minerva,* Vol. XXX, No. 3 (Autumn 1992), pp. 317–336.

Lewis M. Branscomb, "Technology Policy and Economic Competitiveness," *Science and Technology Policy Yearbook 1992* (Washington, D.C.: American Association for the Advancement of Science, 1993)

Harvey Brooks, "Typology of Surprises in Technology, Institutions, and Development," in William C. Clark and R.E. Munn, eds., *Sustainable Development of the Biosphere* (Cambridge, New York: Cambridge University Press, 1986) pp. 325–350

D. Allan Bromley, *The U.S. Technology Policy* (Washington, D.C.: The Executive Office of the President, September 26, 1990)

William J. Clinton and Albert Gore, Jr., *Technology for America's Economic Growth: A New Direction to Build Economic Strength* (Washington, D.C.: The White House, February 22, 1993); see also Clinton-Gore Campaign, *Technology: The Engine of Economic Growth: A National Technology Policy for America* (Little Rock, Ark.: Clinton-Gore Campaign Headquarters, September 21, 1992)

Linda Cohen and Roger Noll, *The Technology Pork Barrel* (Washington, D.C.: The Brookings Institution, 1991)

Committee on Science Engineering and Public Policy, *The Government Role in Civilian Technology: Building a New Alliance* (Washington, D.C.: The National Academy Press, April 1992)

Competitiveness Policy Council, *A Competitiveness Strategy for America*, Second Report to the President and Congress (Washington, D.C.: The Competitiveness Policy Council, March 1, 1993)

Enabling the Future: Linking Science and Technology to Societal Goals (New York: The Carnegie Commission on Science, Technology, and Government, September 1992)

George Heaton, "Commercial Technology Development: A New Paradigm of Public-Private Cooperation," *Business in the Contemporary World*, Autumn 1989, pp. 87-98

Margaret O. Meredith, Stephen D. Nelson, and Albert Teich, eds., *AAAS Science and Technology Yearbook, 1991* (Washington, D.C.: American Association for the Advancement of Science, 1991)

National Academy of Engineering, Committee on Technology Policy Options in a Global Economy, *Mastering a New Role: Shaping Technology Policy for National Economic Performance* (Washington, D.C.: National Academy Press, May 1993)

The National Science and Technology Policy, Organization, and Priorities Act of 1976, 42 U.S. Code 6683

Richard R. Nelson, *National Innovation Systems: A Comparative Analysis* (New York: Oxford University Press, 1993)

Michael E. Porter, *The Competitive Advantage of Nations* (New York: Free Press, 1990)

President's Commission on Industrial Competitiveness, *Global Competition: The New Reality*, Vol. 1 and 2, John Young, chairman (Washington, D.C.: U.S. Government Printing Office, 1985)

Technology and Economic Performance: Organizing the Executive Branch for a Stronger National Technology Base, Admiral Bobby Inman, task force chair (New York: Carnegie Commission on Science, Technology, and Government, 1991)

The Technology Race: Can the U.S. Win? The J. Herbert Hollomon Memorial Symposium (Cambridge: MIT Center for Technology, Policy and Industrial Development, 1991)

Defense Technology and Export Controls

Aerospace Industries Association, *Key Technologies for the 1990's: An Overview* (Washington, D.C.: The Aerospace Industries Association, 1987)

Defense Critical Technologies Plan (Washington, D.C.: The Department of Defense, March 15, 1989); also issued in 1990

Defense Science Board, *Report of the Defense Science Board Task Force on Export of U.S. Technology* (Washington, D.C.: Office of the Secretary of Defense, February 27, 1976)

B. Edelson and R. Stern, *The Operation of DARPA and its Utility as a Model for a Civilian DARPA* (Washington, D.C.: The Johns Hopkins Foreign Policy Institute, November 1989)

Office of the Undersecretary of Defense Acquisition, *The Militarily Critical Technologies List* (Washington, D.C.: Department of Defense, October 1986)

Program Information Package for Defense Technology Conversion, Reinvestment, and Transition Assistance (Washington, D.C.: Advanced Research Projects Agency, March 10, 1993)

U.S. Congress, Office of Technology Assessment, *Holding the Edge: Maintaining the Defense Technology Base*, OTA-ISC-420 (Washington, D.C.: U.S. Government Printing Office, April 1989)

Critical Technologies and Their Public Policy Implications

The Conference Board, *Managing Critical Technologies: What Should the Federal Role Be?* Research Report No. 943 (Washington, D.C.: Conference Board, December 14, 1989)

Council on Competitiveness, *Gaining New Ground: Technology Priorities for America's Future* (Washington, D.C.: Council on Competitiveness, March 1991)

Emerging Technologies: A Survey of Technical and Economic Opportunities (Washington, D.C.: U.S. Department of Commerce Technology Administration, 1990)

Mary Ellen Mogee, *Technology Policy and Critical Technologies: A Summary of Recent Reports*, The Manufacturing Board Discussion Paper No. 3 (Washington, D.C.: The National Academy Press, December 1991)

Perspectives: Success Factors in Critical Technologies (Washington, D.C.: Computer Systems Policy Project, July 1990); and *Perspectives on U.S. Technology Policy, Part II: Increasing Industry Involvement* (Washington, D.C.: Computer Systems Policy Project, February, 1991)

Report of the National Critical Technologies Panel (Washington, D.C.: National Critical Technologies Panel, 1991); see also same title, 1993.

Information Infrastructure

Lewis M. Branscomb, "U.S. Scientific and Technical Information Policy in the Context of a Diffusion-Oriented National Technology Policy," *Government Publications Review*, Vol. 19, No. 5 (September/October 1992), pp. 469–482

Committee on Physical, Mathematical and Engineering Sciences, Federal Coordinating Council for Science, Engineering and Technology, Office of Science and Technology Policy, *Grand Challenges 1993: High Performance Computing and Communications*, a Supplement to the President's Fiscal Year 1993 Budget (Washington, D.C.: The White House, 1992)

Executive Office of the President, Office of Science and Technology Policy, *The National Research and Education Network Program: A Report to Congress* (Washington, D.C.: Executive Office of the President, December 1992)

Brian Kahin, ed., *Building Information Infrastructure* (New York: McGraw-Hill/ Primis, 1992)

National Research and Education Network Review Committee of the Computer Science and Technology Board, Commission on Physical Sciences, Mathematics and Resources, National Research Council, *Toward a National Research Network* (Washington, D.C.: National Academy Press, 1988)

National Telecommunications and Information Administration, *Telecommunications in the Age of Information: The NTIA Infrastructure Report*, NTIA Special Publication 91-26 (Washington, D.C.: U.S. Department of Commerce, October 1991)

U.S. Congress, Public Law 102-94, *High Performance Computing Act of 1991*, Washington, D.C., December 9, 1991

U.S. Congress, Office of Technology Assessment, *Informing the Nation: Federal Information Dissemination in an Electronic Age*, OTA-CIT-396 (Washington, D.C.: U.S. Government Printing Office, October 1988)

Technology Transfer: State and Federal Extension Services

C. Christopher Baughn and Richard N. Osborne, "Strategies for Successful Technological Development," *Technology Transfer,* Vol. 14, No. 3 & No. 4 (Summer–Fall 1989), pp. 5–13

Marianne K. Clark, *A Governor's Guide to Economic Conversion* (Washington, D.C.: Capital Resources Policy Studies; Economic Development, Science and Technology Program; National Governors' Association, 1992)

Stephen S. Cohen and John Zysman, *Manufacturing Matters: The Myth of the Post-Industrial Economy* (New York: Basic Books, 1987)

Maryellen R. Kelley, *Productivity and Information Technology: The Elusive Connection*, Working Paper Series 92-2 (Pittsburgh: School of Urban and Public Affairs, Carnegie Mellon University, January 1992)

W.H. Lambright and A. Teich, "Federal Laboratories and Technology Transfer: An Interorganizational Perspective," in D.E. Cunningham, et al., eds., *Technological Innovation* (Boulder, Colo.: Westview Press, 1977), pp. 425-440

David C. Mowery, "The U.S. National Innovation System: Origins and Prospects for Change," *Research Policy*, No. 21 (Fall 1992)

National Research Council, Manufacturing Studies Board, Committee for the Study of the Causes and Consequences of the Internationalization of U.S. Manufacturing, *The Internationalization of U.S. Manufacturing: Causes and Consequences* (Washington, D.C.: National Academy Press, 1990)

National Research Council, Manufacturing Studies Board, Committee on Analysis of Research Directions and Needs in U.S. Manufacturing, *The Competitive Edge: Research Priorities for U.S. Manufacturing* (Washington, D.C.: National Academy Press, 1991)

Science, Technology and the States in America's Third Century (New York: Carnegie Commission on Science, Technology, and Government, September 1992)

Philip Shapira, *Modernizing Manufacturing: New Policies to Build Industrial Extension Services* (Washington, D.C.: Economic Policy Institute, 1990)

Philip Shapira, J. David Roessner, and Richard Barke, *Federal-State Collaboration in Industrial Modernization* (Atlanta School of Public Policy, Georgia Institute of Technology, July 1992)

Gene R. Simons, Ron Brandow, and Michele Chank, "Technology Transfer in Supplier Development Programs," *Proceedings of the Technology Transfer Society Annual Meeting* (Denver, 1991)

William E. Souder, Ahmed S. Nashar, and Venkatesh Padmanabhan, "A Guide to the Best Technology-Transfer Processes," *Technology Transfer,* Vol. 15, No. 1 & No. 2 (Winter–Spring 1990), pp. 5-16

Strengthening of America Commission (Washington, D.C.: Center for Strategic and International Studies, 1992)

Louis G. Tornatzky and Daniel Luria, "Technology Policies and Programmes in Manufacturing: Toward Coherence and Impact," *International Journal of Technology Management,* Special Issues on Strengthening Corporate and National Competitiveness Through Technology, Vol. 7, No. 1, No. 2, & No. 3 (Spring 1992)

Debra Wince-Smith, "A Vision for Shared Manufacturing," *Mechanical Engineering* (December 1990)

James P. Womack, Daniel T. Jones, and Daniel Roos, *The Machine that Changed the World* (New York: Rawson Associates, 1990)

Universities, NSF, and NIH

Erich Bloch, for the Government-University-Industry-Research Roundtable, *Fateful Choices: The Future of the U.S. Academic Research Enterprise: A Discussion Paper* (Washington, D.C.: National Academy Press, March 1992)

Harvey Brooks, *The Research University: Centerpiece of Science Policy?* Working Paper WP 86-120 (Columbus: The Ohio State University, College of Business, December 1986)

Wesley Cohen, Richard Florida, and W. Richard Goe, *University-Industry Research Centers in the United States: A Report to the Ford Foundation* (Pittsburgh: Center for Economic Development, Carnegie Mellon University, June 1992)

Government-University-Industry Research Roundtable, *The Future of the Academic Research Enterprise: Report of a Conference*, James D. Ebert, Chair (Washington, D.C.: National Academy Press, 1992)

In the National Interest: The Federal Government and Research-Intensive Universities, A Report to the Federal Coordinating Council for Science, Engineering, and Technology (FCCSET) from the Ad Hoc Working Group on Research-Intensive Universities and the Federal Government (Washington, D.C.: Office of Science and Technology Policy, December 1992)

National Science Board Commission on the Future of the National Science Foundation, *A Foundation for the 21st Century: A Progressive Framework for the National Science Foundation* (Washington, D.C.: The National Science Board, November 20, 1992)

Office of Technology Assessment, *Federally Funded Research: Decisions for a Decade*, OTA-SET-490, Daryl E. Chubin, project director (Washington, D.C.: U.S. Congress, Office of Technology Assessment, May 1991)

President's Council of Advisors on Science and Technology (PCAST), *Renewing the Promise: Research-Intensive Universities and the Nation* (Washington, D.C.: U.S. Government Printing Office, December 1992)

Human Resources

Lewis M. Branscomb, chair, Task Force on Teaching as a Profession, *A Nation Prepared: Teachers for the 21st Century* (Washington, D.C.: Carnegie Forum on Education and the Economy, May 1986)

Harvey Brooks, "Educating and Training the U.S. Workforce for the Twenty-first Century," in Dorothy S. Zinberg, ed., *The Changing University: How Increased Demand for Scientists and Technology is Transforming Academic Institutions Internationally* (Dordrecht, The Netherlands: Kluwer Academic Publishers, 1991)

Carnegie Commission on Science, Technology and Government, *In the National Interest: The Federal Government in the Reform of K-12 Math and Science Education*, Lewis M. Branscomb, task force chair (New York: Carnegie Commission in Science, Technology, and Government, September 16, 1991)

Commission on Skills of the American Workforce, *America's Choice: High Skills or Low Wages!* (Rochester, NY: National Center on Education and the Economy, 1990)

Richard M. Cyert and David C. Mowery, eds., *Technology and Employment: Innovation and Growth in the U.S. Economy* (Washington, D.C.: National Academy Press, 1988)

The Forgotten Half: Pathways to Success for America's Youth and Young Families (Washington, D.C.: The William T. Grant Foundation Commission on Work, Family, and Citizenship, 1988)

William B. Johnston and Arnold E. Packer, *Workforce 2000: Work and Workers for the 21st Century* (Indianapolis: Hudson Institute, 1987)

Richard Kazis, *Improving the Transition from School to Work in the United States* (Cambridge, Mass.: Jobs for the Future, 1991)

Jonathan Kozol, *Savage Inequalities: Children in America's Schools* (New York: HarperCollins, 1991)

Jonathan Kozol, *Illiterate America* (New York: Anchor Press/Doubleday, 1986)

Ray Marshall and Marc Tucker, *Thinking for a Living: Education and the Wealth of Nations* (New York: Basic Books, 1992)

Robert B. Reich, *The Work of Nations: Preparing Ourselves for 21st Century Capitalism* (New York: Alfred A. Knopf, 1991)

Index